Place of Science
in a World of
Values and Facts

INNOVATIONS IN SCIENCE EDUCATION AND TECHNOLOGY

Series Editor:
Karen C. Cohen, Harvard University, Cambridge, Massachusetts

A Continuation Order Plan is available for this series. A continuation order will bring delivery of each new volume immediately upon publication. Volumes are billed only upon actual shipment. For further information please contact the publisher.

Place of Science in a World of Values and Facts

Loucas G. Christophorou

Kluwer Academic / Plenum Publishers
New York, Boston, Dordrecht, London, Moscow

Library of Congress Cataloging-in-Publication Data

Christophorou, L. G.
   Place of science in a world of values and facts/Loucas G. Christophorou.
     p. cm. — (Innovations in science education and technology)
   Includes bibliographical references and index.
   ISBN 0-306-46580-9
     1. Science—Philosophy. 2. Science—Social aspects. 3. Science—Methodology. I. Title.
II. Series.

Q175 .C4766 2001
501—dc21

2001016496

ISBN: 0-306-46580-9

©2001 Kluwer Academic/Plenum Publishers, New York
233 Spring Street, New York, New York 10013

http://www.wkap.nl

10 9 8 7 6 5 4 3 2 1

A C.I.P. record for this book is available from the Library of Congress

Printed in the United States of America

TO MY FAMILY
with Gratitude

Foreword

This is an engrossing book. It is also an unusual book: it is written by a scientist who is quite willing to talk about the softer side of life, about things such as love and respect and responsibility, and to try and position them in the context of his science. He is also willing to talk about religion, the manner in which it relates to science and science to it, and to attempt reconciliation of both. He sets himself a tough task, to tread the narrow path between the maudlin and the severely sober. In this, he is eminently successful. He is successful not because he aims at any grand synthesis, but because he has chosen the more modest path of simply laying out the cards on the table.

This work is also unusual for another reason. The majority of books that attempt to explain science to a lay public, that try to describe its workings, its *raison d'être*, its hidden contents, its societal impact, its implications for our future, etc., are written by theorists. This is hardly surprising. The theoretician, after all, is expected to think deeply, to be the great unifier, to be concerned with meaning. Very few books about science are written by scientists, ones who spend their time in a working experimental laboratory. This is such a book. And because it is, it is also a very different book. It comes from a scientist who runs a working laboratory, employs scientists, educates students, solicits grant monies…in short, a person who is totally immersed in reality. The author is familiar with the frustrations of instrumentation that should work but does not, with artifact that stubbornly refuses to go away, with administrators who expect accountability to an irrelevant or even deleterious set of norms, with grantor sources that want to "steer" the research or deny funding because they can't, and with constant demands for a spurious "utilitarianism." It is a pleasure to run across a book that carries the imprimatur of a laborer in the science vineyard, the laboratory. This book, then, is very different, and it is a gem.

This is a cautiously optimistic book. As the author says: "I, too, 'like the dreams of the future better than the history of the past,' but I also like the beacons of inextinguishable hope of the past, the Phoenix of the Greeks and the Easter of Christianity." In the process, he advocates a neo-teleogical philosophy, one that does not so much presume the inevitability of sociological progress as it does the critical need for development of a humanities/science interface that will minimize sociological disaster. He wants to imbue each of them, the humanities and the sciences, with the values of the other: to have the provisionality of science moderate the absolutism of religion, and to make science face up to the fact that the humanities and the religions can fill the void that now lies outside its purviews. It is this conjunction, he believes, that will realize the dream of a "hopeful future where everything that can go well will go well and everything that may go wrong can and will be prevented."

Professor Christophorou gives an excellent discussion of the "values" inherent in science. We will mention only one of the instigators of these values, namely "transience." Everything of science, its ideas, theories, laws, even facts... everything, that is, except the method of science... all are transient. Science, you see, cannot prove anything. However, it is exceedingly well equipped with the tools required for disproof. Thus, since nothing can be proven, the greatest continuous string of successful experimental verifications will be totally interrupted by just one contradictory fact, and the prior ideology will have to be revised, reconstructed or as necessary discarded in order to satisfy that single obstruent fact. This has happened time and again. No doubt, it will recur, and these happenings will carry with them all the preconditioning for values such as humility, tolerance and patience: humility, because one could be wrong; tolerance, because the other fellow could be right; and patience, because a process exists that will eventually work it all out.

This book is concerned with the gap between where we are and where we want to be. In his attempt to bridge that gap, the author becomes many things: brave, adventurous, prescient, vulnerable, etc. But he is never oblique, and he is always entertaining. He has written a unique and valuable book that I can enthusiastically recommend to any reader.

Sean P. McGlynn
Baton Rouge

Preface

This book is about science and beyond. It looks at science and its place in society holistically and comprehensively. The gap between those in science and those outside science is bridged by relating broad social and philosophical issues to science, and by connecting science and its method to human functions and the condition of modern society.

The book is intended for the student, teacher, and worker in science, and for those non-scientists who agonize about science and technology, the values of man, and the future of both science and society.

The material has been organized coherently: each chapter flows from those preceding it. Chapter 1 is an overview of man in the cosmos. Chapter 2 looks at the societies of modern man and their most distinct characteristics. Both chapters set the stage for understanding the role of science in a world of values and facts. Chapter 3 is a condensed account of the evolution of science -- mostly physics -- and aims at understanding how science works, evolves, and impacts man. Chapter 4 focuses on selected recent scientific advances demonstrating the ability of science to penetrate the physical universe and to establish the laws that describe its observed behavior. Chapter 5 concentrates on distinct characteristics and principles of science which define its tradition and qualify its functions. Chapter 6 focuses on the people who work in and do science and makes a distinction between the scientist and the science worker. The technological value of basic scientific research and its connection to technology is documented in Chapter 7. Chapter 8 looks at the cultural and educational value of science and Chapter 9 attempts to show the complementarity of science and religion and the need to jointly address social and ethical issues brought about by science which transcend both science and religion. In Chapter 10 are discussed natural and socio-political limits of science. The last Chapter is an outlook on the future of and in science.

It is hoped that the process and the medium the book describes will give form to knowledge that will complement the reader's vision of the place of science in a world permeated by values and dominated facts.

I am grateful to many friends and colleagues for innumerable discussions and encouragement, and to my former students at the University of Tennessee with whom I shared parts of the material covered in these pages. Thanks are due to Professor Benjamin Bederson who went through the manuscript and made valuable comments and suggestions, and to Professor Sean P. McGlynn who wrote the book's Foreword. To my wife *Eratoula* and daughters *Penelope* and *Yianna* I owe a deep debt of gratitude for their support and for shaping my perspective.

                                         LOUCAS G. CHRISTOPHOROU

Germantown, Maryland

Contents

CHAPTER 1

The World
of
Values and Facts

1.1. US AND THE COSMOS

Man's search for a place in the cosmos is perennial. Deep in it lies the meaning of his own existence. Since the idea of the cosmos was first conceived by the Greeks of antiquity, man (the word man is used throughout the book to mean humankind) realized that to understand himself he needs to comprehend the universe he lives in and is a part of. It appears that life is more meaningful that way. In this eternal quest science has become the key for opening the windows of the cosmos.

The cosmos -- with its incomprehensible size and age, large amounts of luminous and even larger amounts of non-luminous (dark) matter, strange forces, and its "prodigious emptiness"[1] -- is overwhelming. Its origin is mysterious but stretches something like 30 billion light years in the past.[2] It is a restless cosmos filled with billions of galaxies which evolve under the force of gravity. It is a universe full of radiation, particles, and gravitational forces. A ceaseless cosmic dance of the magnificent transformations of energy.

Our home, the planet earth, is a tiny and insignificant speck in the macrocosmos. We have, however, seen our tiny cloud-wrapped "blue" planet from a distance (Figure 1.1) and realized its preciousness as the only planet we can call home. It is more than four billion years old and life on it traces back to something like 500 million years. The human family emerged on this planet about seven million years ago, and our ancestors originated in Africa something like 200,000 years back.[3] Human cities and recorded history are relatively newer events: they barely began five thousands years ago or so.

Figure 1.1. The Earth rising over the Moon as seen by the Apollo 8 astronauts as they became the first humans to circumnavigate the Moon (Courtesy of NASA).

Compared with these dimensions of space and time we are not much to speak of: our size is too small and our life span is too short. This measure is, however, relative. It changes greatly when we enter the microcosmos, the realm of atoms, nuclei, elementary particles, and the reactions of the constituents of the microcosmic nature. Here sizes are incredibly small and lifetimes of events and entities can be extraordinarily short. The diameters of the atoms are as small as 10^{-10} to 10^{-9} meters and their masses as little as 10^{-27} to 10^{-25} kilograms. Their mass resides almost entirely in an even smaller region in their center, the nucleus, the rest of the atomic volume being essentially empty. And yet, as far as we now know, everything in this gigantic universe (the galaxies, the planets, the air, the water, the plants, and all living organisms) is made up of the same minute particles, the atoms and their constituents. It takes something like 10^{28} of these atoms to build up a human body, and about 2.4×10^{36} diameters of the hydrogen atom to make up the length of the universe, about 2.4×10^{23} kilometers.

Similarly, the lifetimes of the atomic and subatomic entities and the duration of microcosmic events are extraordinarily short. Atoms, for instance, remain excited normally for about 10^{-9} to 10^{-8} seconds. A number of atomic species live, and many atomic phenomena occur, for only 10^{-15} to 10^{-13} seconds. Some elementary particles exist for yet shorter times, about 10^{-23} seconds. If we take the average lifetime of man to be about one-half century, we see that man lives about 10^{17} times longer than excited atoms do.

Compared to the sizes of the microcosmos, then, man appears to be a giant, and compared to the time scales of the microcosmic events, then, man appears to be eternal.

The activity of science spreads over the entire range of these incredible dimensions of size and time (Figures 1.2 and 1.3). The scientific frontier stretches from the cosmic to the atomic, from the ends of the universe to the innermost regions of the most microcosmic entities in the center of the atom, from the Big Bang (the instant of cosmic expansion) some twenty billion years ago, to the shortest-lived elementary particles, over distances and times varying by a factor greater than 10^{40}.

A common characteristic of the microcosmic and the macrocosmic universes is their enormous emptiness. Matter and life, it has been noted,[1] are contained between these two voids.

Humanity experienced the macrocosmos first and for almost its entire history. In contrast, our experience with the microcosmos is recent. Both experiences are still largely outside the culture and appreciation of the "lay" person. It is as difficult for the non-specialized physicist, for instance, to comprehend the nature of the nuclear forces, as is to understand the

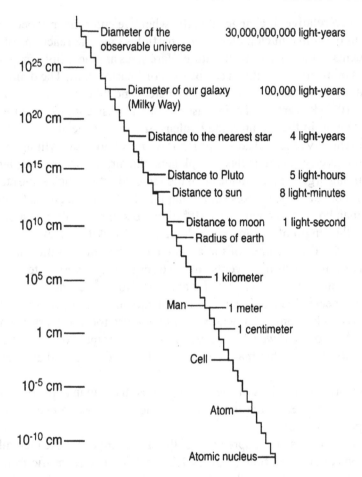

Figure 1.2. Range and sizes in the macrocosmos and microcosmos (each step corresponds to a factor of 10 increase in size) (from Ref. 4).

meaning of the billions of light years of the macrocosmic events, or to meaningfully contemplate the nature of the universe as early as 10^{-33} seconds after the "Big Bang."[5] And yet, this is the beauty and the marvel of science: the incomprehensible extension of man's "vision."

There is no evidence that life as we know it on earth exists in any of the billions of stars which surround us, no matter how strange this may appear to be. The key element in the evolution of the universe and prebiotic molecules -- carbon --, and many molecules believed to be involved in basic life from methylidyne (CH) and cyanogen (CN) radicals, to water vapor

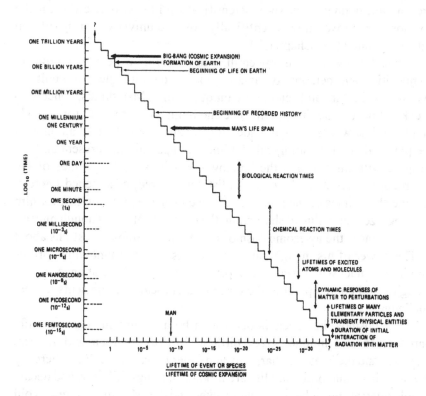

Figure 1. 3. Range and lifetimes of events or species plotted as the logarithm of time versus the ratio of the lifetime of event or species divided by the lifetime of cosmic expansion (each step on the vertical axis corresponds to a factor of 10 increase in time).

(H_2O) and formaldehyde (H_2CO), to formic acid (HCOOH), acetaldehyde (CH_3CHO) and ethyl alcohol (C_2H_5OH), are now known to exist in interstellar space;[6,7] amino acids have been found imbedded in meteorites;[7] and fossilized "nanobacteria" are claimed[8] to have been observed in a rock presumed to have fallen on earth from Mars. Does this indicate that life is scattered across the universe? We simply do not know. Life on earth is intimately bound by delicate conditions such as temperature (we cannot live on the sun), forces of gravity (we cannot live at the bottom of the oceans), and surface environment (we cannot live on surfaces bombarded by intense radiation). That said, life might exist elsewhere in the universe under entirely different set of conditions. Even the earth, when looked at from the planets, seemed a hostile environment, full of metallic vapors. At the present time neither do we know scientifically if life exists elsewhere in the cosmos, nor do we know scientifically how the universe and the life on earth originated (see Chapter 10).

Man, however, exists on earth, and exists as a person, unique in composition and personality, as a product of the biological and cultural evolutions. Culture and science are unique to him. Man thinks, conceives, speaks, chooses, uses his intelligence and brain power to work for him. He laughs and he weeps. He has a sociological and a moral world along with the physical and the biological. To him belongs only the present and the struggle with himself and the unknown. For him the long times of the macrocosmic events make them indistinct, blurred, of lessened interest, and the short times of the microcosmic events make them too fast for him to recognize their value in sharpening the contrasts of life. Though himself transient, faces the awesome cosmos asking innumerable questions about it. The power of his thought and comprehension outdistances time, brings him into contact with distant stars, places him everywhere at once. To his thoughts and contemplation, the cosmos, as Pascal said, presents no real challenge.

The cosmos, however, is permeated by a continuous change. Life, unrehearsed and unique, gives birth to life, trees green and flowers blossom, and the beauty of harmony and order radiate hope. Elsewhere, by the same laws that order and life are produced, things fall apart and decay, people quarrel and fight, loved ones sicken and die. Man himself grows old and becomes increasingly cognizant of the distinct advantage the cosmos has over him -- virtually all of humanity is replaced every 50 or so years. Nevertheless, we recognize, as Pascal did, that "of all this the physical universe knows nothing."[9] Herewith, then, rests the realization of man's smallness and man's bigness. A realization that human existence is not merely an interplay of physical forces to be understood by science alone, but a miracle that transcends science, requiring illumination by the complementary forces from within.

1.2. A FEW THOUGHTS ON HUMAN HISTORY

Through recorded history,[10] generation after generation inherits knowledge and skills, culture and values, and a particular perception of the world. Through it we learn that civilizations rise, flourish, and fall. The annals of history reveal that nations prosper when they are spiritually and intellectually strong, and break down when they fail to plan and adapt, or when they are complacent, or, worse yet, when they willfully abdicate their freedom. History teaches that societies, and the eras to which they attach their names, are recognized by their central value systems, the wisdom by which they define their priorities, the manner in which they care for their underprivileged members, and the way in which they treat their environments. Throughout history, observed Dubois,[11] societies concerned themselves much more with the development and maintenance of institutions than with the welfare of their members. It can perhaps be argued as well that past and present societies have succeeded in providing a dignified life for only a small fraction of their members. Most of the older civilizations were male and slave dominated, and a large fraction of humanity still awaits emancipation! Science and science-based technology can change these conditions. As we shall see in subsequent chapters, science has proven Malthusian predictions wrong, made Darwinism inapplicable to the modern human, and penetrated the iron curtains of totalitarian states more effectively than the mightiest armies human history speaks of.

The rise and the fall of civilizations is paralleled with that of innovating societies. Both had been transient, both flourished in local environments, and both constituted the exception rather than the rule: the Far Eastern civilizations flourished in Northern China; those in the Americas before Columbus in Peru and Mexico; the prehistoric and the classical ones between about 3,000 B. C. and 200 A. D. were located in Egypt, Asia Minor, Cyprus, and Greece; the more recent (14th and 18th century) centered in England, France, Iberia, and Germany. Always man excelled in relatively small settings and only for a short period of time. China, Mesopotamia, Egypt, and Greece were once great innovating centers, but subsequently ceased innovating for centuries. It seems that not only appropriate economic, political, and social conditions are required for innovation, but also a free spirit, an open mind, and a dedicated attitude toward a better future are necessary to kindle and to sustain the flame of innovation. Flames are nonetheless temporal and ephemeral. They are always condemned to extinction. Human history tells us however that their luminosity and majestic beauty rekindle new flames in due course. The spark of innovation that was extinguished long ago in the Far East and the Eastern Mediterranean, rekindled centuries later the flames in the West.

Human history distinctly teaches that mankind lived dangerously, and in a continuous state of transition. Man engaged himself in an endless succession of wars which he fought at every level from the tribe and chiefdom, to the city and nation state, to the empire and block of nations, to the world. Wars paraded under colorful slogans during Pax Romana, Pax Britannica, Pax Americana, and Pax Sovietica. They were only infrequently interrupted by short periods of peace. Wars were proclaimed in the name of morality and absolutism and were fought by "the cheated many for the cheating few." They were caused by greed, thirst for power, or sheer irrationality. Others had their roots in despair, fear, or ignorance, in irreconcilable social differences, incompatible human variables, or the clash of incoherent social structures out-of-phase with each other. Yet others, the revolutions, were inspired by the dedicated few on behalf of the oppressed many. Without exception the uncontrolled fires of wars destroyed the green and the tender and silenced the laughter and the hope of men. Even the revolutions often freed man from one evil and trapped him by another.

Wars have been fought with every murderous and destructive means man's ingenuity has ever conceived, history tells us. Whether at the time of Archimedes or at the time of Newton or at the time of Einstein, the weapons of war were developed through science and technology. The frontiers of science and technology had been then and continue to be today the frontiers of weaponry (Chapters 2 and 10).

Turning points in human history are also distinct changes in man's thinking, in man's attitude toward his fellowman, and in man's approach toward nature. The discovery of the deductive method and the application of the inductive inference in classical Greece, and the invention of the experimental method in Renaissance Europe are examples of the first. The Christian ethic of justice and love and the Roman legal tradition are examples of the second. Man's ingenious ways of controlling, extracting, and using energy and his recent attitude toward the environment are examples of the third.

Turning points in human history are also the works of man the builder, man the creator, man the inventor, and man the intellectual revolutionary. Man has built the pyramids and the Parthenon, composed symphonies, and fathomed Pietas. Man has constructed intellectual and verbal systems through which he communicates his finest thoughts and his innermost feelings. Man has invented writing, the wheel, the compass, and the techniques of handling bronze and iron. Man harnessed the wind, communicated through printing and across the vacuum, learned to burn rocks for energy, and traveled through outer space.

Paralleling the monotonic increase of man's knowledge is the increase of his social units. The latter evolved from the village and the chiefdom, to the city-state and the nation-state, to the blocks of nation states, and to the United Nations. Human societies underwent a gradual transition from agricultural to industrial and now -- in certain parts of the world -- to post-industrial stages. Man's social organizations continuously moved toward greater level of complexity. This "seemingly inexorable rise in the level of complexity is driven by cultural evolution not by biological change," stressed Leakey, for biologically we are the same people as our ancestors 7,000 years ago except that they were without the infrastructure of civilization.[12] But perhaps the most important teachings of human history are those which relate to the meaning of happiness and greatness; dignity and honesty; gratitude and generosity; freedom and truth; justice and love (αγάπη). Those old and enduring values which are associated with the realization that there is a God as well as a devil within each person. These concepts and values have occupied the thoughts and the hearts of the finest men and women throughout history. They varied in color and emphasis both with the time and with the place. Their evolution gave birth to characteristic sets of traditional value systems which constituted *the frames of reference for value judgement* of the many microcultures that inhabited and inhabit the earth.

These values and the distinct teachings of the turbulent human past must be understood warned Santayama, because those who do not understand history are destined to repeat its mistakes. They must also be understood, instructed others, because those who were truly great remembered the souls of history. It is important, therefore, that the facts of history -- as the facts of science -- remain as such: facts. The deliberate distortion of historical facts is a crime against humanity. The human past is immanent in humanity's present and in humanity's future. No set of values is meaningful without the meaning of human history. Both the past and the future are affected by and affect science.

1.3. THE MICRO (MULTI) CULTURES OF THE WORLD

There are on earth many worlds, wrote Renè Dubois,[13] and we agree. There are on earth many microcultures normally connected with national characteristics tracing back centuries and identified with specific motherlands. The richness and the multiplicity of these cultures were always spread unevenly over the face of the earth. They are a strong manifestation of the magnificent diversity of the human race and the *genius loci.*

Interwoven in the fabric of each microculture are its inherent constants: old and strong traditions, distinct cumulative aspects of a specific way of life with written and unwritten norms, myths, history, faith, sacred places, and ancestry. Associated with each microculture is a value system which qualifies, enriches, and limits the lives of its people and provides a rational basis for their view points and value judgements. Within a microculture different but compatible individuals, constituting the nations (ἔθνη) of the world, can live and function comfortably. In each such ethnic group noted Spicer,[14] there is an intense collective consciousness and a high degree of internal solidarity, a mutually shared system of values that accounts for internal stability and conveys a sense of security. Usually, there is also a common language[15] which instills pride and affords a shield from other traditions and value-judgement systems perceived to be harmful. Man's culture speaks in many languages indeed. Culture, wrote Bloom,

> "restores the lost wholeness of first man on a higher level, where his faculties can be fully developed without contradiction between the desires of nature and the moral imperatives of his social life ..... Culture is almost identical to people or a nation, ... it refers to art, music, literature, everything that is uplifting and edifying .... it is what makes possible the rich social life that constitutes a people, their customs, styles, tastes, festivals, rituals, gods .... it generates its own way of life and principles, particularly its highest ones, with no authority above it."[16]

Diversities in ideology and cultural viewpoints, although an additional manifestation of the richness of human nature, are, nonetheless, sources of resistance to social change and at the root of many human conflicts. Each microculture fights to hold to its own and to claim a place for itself on this planet.

Each microculture affects and is affected by modern science. Each can be penetrated from without and from within by the means provided by science, independently of the microculture's will or involvement in the scientific enterprise. Yet, microcultural outlooks have been shaped long before modern science. This is largely the reason why at the dawn of the 21rst century when the world is unified in so many ways by science and technology (Chapter 2), the microcultures of the world still maintain fundamentally different outlooks. The miracle of life as is expressed in the diversity of the microcultures has majestic beauty and magnetic attraction. The microcultures that man has created, create him. They unite mankind and they separate mankind. They preserve the richness of mankind and they often perpetuate the ugliness of chauvinism. While there is still a great deal of validity in the Delphic Oracle's answer to him who asked how it was proper to serve God that he should go according to the laws and customs of (his) country, every nation needs to look itself in the mirror of human history and ponder over its own image.

How can the microcultures of the world be maintained as *complementary*[17] aspects of a single human world culture, the one that stresses the single identity and family of man, a family shaped by eons of common struggles and "awe at the miracle of life?"[18] How can the implicitness of the microculture be harmonized with the transcendence of the human world culture? Undoubtedly, this is and will remain a heterogeneous world, one that demands preservation of national characteristics and discreetness toward them. However, the demand for differentiation need not oppose the demand for symbiosis. The rights of a citizen in a nation state need not be incompatible with the rights of a human being anywhere. An equilibrated and coupled world should not be a structureless society.

The unification of the world microcultures requires *their mutual convergence to a common frame of reference for value judgement*. This is the most important single condition for the symbiosis of the peoples of the world. There is a need for a mutual convergence of the microcultural value judgement systems of the world to a world-wide value-judgement system that will transcend, but never replace the world's microcultural value judgement systems. A convergence based on fundamental elements common to the world microcultures, their positive attributes and truths, and the unchangeable and transnational characteristics of man, the truly human, the timeless.

1.4. THE TIMELESS

Ruth Anshen, in her preface to the book *Science: The Center of Culture*,[19] posed a fundamental question and provided an equally fundamental answer to it. She asked: "What is it that endures and is the foundation of our intellectual and moral civilization? What is it that must survive and be transmitted to the future if man is to remain human?" Her vivid answer was: "It is that heritage of timeless and immutable values on which we can fix our gaze whenever the language of change and decline which history speaks seems to become too overwhelming for the human heart."

The immutable values Anshen speaks of are the standards by which we judge the significance of life. Those aspects of human life that transcend generations and microcultures. Those deeper illuminations in the light of which justice and injustice, freedom and slavery, and good and evil are in sharp contrast. Those all inclusive concepts which -- while vaguely defined and often seemingly irrational -- are with objective validity, and constitute a storage of wisdom and a basis for private convictions. Those realms which have youthful freshness even though they have been with man for centuries qualifying his virtues, his honesty and honor, his friendliness and generosity, his affection and devotion, his dignity and pride. Those written and unwritten norms of ethical conduct which clearly recognize right and wrong and go beyond justice to embrace the love and the sacrifice of one person for another, and the intimate relations and responses of man which qualify his happiness. The immutable values give cohesiveness to human societies, are "life-preserving and life-enhancing,"[20] are "sharable,"[21] and inspire a hopeful future.

While values underwent many metamorphoses and in a fluid world they are constantly renewed, they remain in their essence invariant, guiding man in his struggle to draw the golden line between that which he wants and that which he may not want but must do, his rights and his duties, for he lives not alone. They force man to be civil by guiding his judgements to achieve that precious balance between his complementary needs as a *free individual person* and as a *social being*.

Thus, in a world dominated by facts there has also been a place for value. What has been lacking, is a unity in the world of values, a value system of wider acceptance, a "value economy."[22] Such unity cannot be based on utility alone. It must rather be based on the demonstrated likeness among the values of the peoples of the world and on the realization that most of the seemingly conflicting values are but complementary. The complementarity[17] I speak of here allows for accommodation between the values referring to the individual person as opposed to those referring to society at large, permits us to cherish our uniqueness as persons and simultaneously to take pride in society's collective achievements, and finds room for the microcultures to preserve their national heritage and simultaneously to be proud members of the world community. It allows for mutual accommodation of concurrently held but differing national and international (transnational) goals. A commonality of conscience founded on commitment which "values the values and makes them valuable."[21]

Unity demands fulfilling man's fundamental obligation to the truth. This obligation needs to be total and unconditional. There is no room for "political correctness." The obligation to the truth requires a search for the truth and a duty to tell the truth. For truth is connected with knowledge and knowledge with freedom, and freedom is the oxygen of society. "The study of man," instructed Polanyi,[23] "must start with an appreciation of man in the act of making responsible decisions," and how can a man make responsible decisions if he is not free? In the modern world it becomes increasingly difficult for a free man to exercise his freedom responsibly without the aid of science and science-based technology.

Finally, human history speaks of transcendental principles of universal validity underneath the multicultural value systems of the world on which a common frame of reference for value judgement can be founded. A bedrock of such fundamental principles as the *Principle of Implicitness*, the *Principle of Reciprocity*, and the *Principle of Love*.

The Principle of Implicitness recognizes the fact that we are never isolated, never privileged observers outside of the system we live in and are a part of. We, therefore, act implicitly. Our actions are embedded within those of our fellowmen. Out of this implicitness stems the need for respect, the necessity for accommodation, and the emergence of the concept of brotherhood. What holds for our actions also holds for our values. Our values' value is implicit. Values owe their value to the existence of other values. This embeddedness of (individual, institutional, national, and human) values allows for their mutual interaction and feedback, their mutual accommodation and indebtedness. *It demonstrates the existence of an underlined unity among human values.* We shall see in Chapter 5 that this is a fundamental principle of science as well because the unity of knowledge is one of the fundamental goals of science.

The Principle of Reciprocity qualifies the supreme value of justice. Do not do to others what you do not like others do to you, proclaimed the Greeks. Do for others everything you want them to do for you, positively instructed Christ.[24] We, thus, decide on good not in an absolute and remote manner, but from the standpoint and in relation to neighboring settings. By nature, we are ethically reciprocative. Ethical reciprocity induces discreetness and ethical sensitivity.

The Principle of Love is the highest principle of all. It is all inclusive and transcendent. It is best described by the Christian tradition. Christ defined its limits as the sacrifice of one's life for his fellowman,[25] Saint John identified[26] it with God, and Saint Paul's description of the Christian Love (Αγάπη) in his first epistle to the Corinthians[27] sounds more like poetry than a command. He wrote:[28]

> "If I speak with the tongues of men and of angels, but do not have Αγάπη, I have become a noisy gong or a clanging cymbal. And if I have the gift of prophecy and know all mysteries and all knowledge, and if I have all faith, so as to remove mountains, but do not have Αγάπη, I am nothing. And if I give all my possessions to feed the poor, and if I deliver my body to be burned, but do not have Αγάπη, it profits me nothing. Αγάπη is patient, Αγάπη is kind and is not jealous; Αγάπη does not brag and is not arrogant, does not act unbecomingly; it does not seek its own, is not provoked, does not take into account a wrong suffered, does not rejoice in unrighteousness, but rejoices with the truth; bears all things, believes all things, hopes all things, endures all things. Αγάπη never fails ..."

The principles just mentioned are pristine in their fundamental universality. Other, more specific, principles can follow from them which we term *derivative principles*. Two examples of such derivative principles can be mentioned. The first is the *Principle of Minimization of Human Suffering*. It was introduced by the scientist Linus Pauling[30] who realized that modern times are *still* a period with too much suffering. It is a call for a coordinated all-out war on the poverty and hunger and the many material causes of suffering that beset the largest fraction of the world *today*. Pauling's principle can be extended to include man's hunger for the timeless and man's need for the immutable values we spoke of in this chapter. We, therefore, wish to refer to Pauling's principle as the *Principle of Maximization of Human Happiness*.

The second derivative principle we wish to refer to is the *Principle of Mutual Trust*. It was introduced[30] by Pope John XXIII who realized the holocaustal dangers of the doctrine of "peace by deterrence." He wrote:

> "All must realize that there is no hope of putting an end to the building up of armaments, nor of reducing the present stocks, nor, still less, of abolishing them altogether, unless the process is complete and thorough and unless it proceeds from inner conviction: unless, that is, everyone sincerely cooperates to banish the fear and anxious expectation of war with which men are oppressed. If this is to come about, the fundamental

> principle on which our present peace depends must be replaced by another, which declares that the true and solid peace of nations consists not in equality of arms but in mutual trust alone..."

In modern times, change and the rate of change have changed the world. Some of these changes we shall see in the next chapter where we focus on modern man and the state of his societies. Even in the modern world with moral issues deeply embedded in the complex structure of science and technology, humaneness cannot be understood without the timeless, and a peaceful world cannot be had by reducing man "to something less than human."[31] The immutable values and timeless truths refer to the permanent rather than to the temporal order, transcend generations and professions, and expand in all directions all the time. They guide scientists and non-scientists alike. They provide a precious continuity in human history. Only man can preserve them.

1.5. REFERENCES AND NOTES

1. C. N. Martin, *The Role of Perception in Science*, Hutchinson & Co. (Publishers) Ltd., London, 1963, p. 37.
2. One light year is the distance traveled by light in one year, i.e., 9.46×10^{12} kilometers. In vacuum (empty space) light travels at a constant speed of about 3×10^{10} centimeters per second. The notation 10^{10} means 1 followed by ten zeroes (i.e., it represents the number 10,000,000,000) and the notation 10^{-10} means one divided by 10^{10} (i.e., it represents the number 0.0000000001 or 1 / 10000000000). In Appendix A are listed various prefixes, their symbols, and meanings.
3. R. Leakey, *The Origin of Humankind*, Basic Books, A Division of Harper Collins Publishers, Inc., New York, 1994.
4. National Academy of Sciences, *Physics in Perspective*, Washington, D. C., Vol. I, 1972, p. 84.
5. F. Close, *The Cosmic Onion*, American Institute of Physics, New York, 1986.
6. D. M. Rank, C. H. Townes, and W. J. Welch, Science **174**, 10 December 1971, p. 1083; Th. Henning and F. Salama, Science **282**, 18 December 1998, p. 2204.
7. F. Hoyle, in *The Greatest Adventure*, E. H. Kone and H. J. Jordan (Eds.), The Rockefeller University Press, New York, 1974, Chapter 1.
8. Science **273**, 16 August 1996, p. 864; Science **276**, 4 April 1997, p. 30; Science **282**, 20 November 1998, p. 1398.
9. A. Krailsheimer, *Pascal*, Oxford University Press, Oxford, 1980, p. 56.
10. A perspective on the meaning of human history seems necessary for the proper appreciation of the place of science in a world permeated by values and dominated by facts. Unfortunately, such a perspective is difficult for many reasons not the least of which because human history is too involved, too subjective, and too contradictory to be easily compressed. The objectivity of human history is also often doubted on the

premise that it has been written by, under the influence of, and about the victors, the powerful, and the rich. Human history may in fact appear to be more optimistic than it had actually been. What has survived or has been recorded through the eyes of the few may be atypical of the people's lives and unrepresentative of their condition and aspirations. In this regard, modern times are better because they are more representative of the totality of the human race and condition, and this improvement has largely been brought about by science and science-based technology.

11. R. Dubois, *A God Within*, Charles Scribner's Sons, New York, 1972, p. 203.

12. Ref. 3, p. 79.

13. Ref. 11, p. 27.

14. E. H. Spicer, Science **174**, 19 November 1971, p. 795.

15. Kroeber (A. L. Kroeber, *Anthropology*, Harcourt, Brace, New York, 1948, pp. 509-537) emphasized the *alphabet* as a classic example of recognizable cultural entity which persisted for centuries. As Spicer[14] put it, "It (the alphabet) existed in recognizably similar form among Phoenicians, Hebrews, Greeks, Arabs, Romans, and so on, into the late 20th century when it has become, in the same ancient, basic form tightly integrated into such knowledge storage and retrieval systems as indexes, encyclopedias, handbooks, city directories, and telephone books, systems on which complex modern societies depend for the maintenance of their total culture patterns."

16. A. Bloom, *The Closing of the American Mind*, Simon and Schuster, New York, 1987, pp.185-193.

17. The concept of complementarity is discussed in Chapters 3 and 5.

18. Ref. 3, p. 157.

19. I. I. Rabi, *Science: The Center of Culture*, World Publishing Co., New York, 1971.

20. Ref. 16, p. 201.

21. T. T. Lafferty, *Nature and Values*, University of South Carolina Press, Columbia, South Carolina, 1976, p. 268.

22. Ref. 21, p. 228.

23. M. Polanyi, *The Study of Man*, The University of Chicago Press, Chicago, Illinois, 1959, p. 71.

24. Saint Matthew, Chapter 7, verse 12.

25. Saint John, Chapter 15, verse 13.

26. First Epistle of John, Chapter 4, verse 8.

27. First Corinthians, Chapter 13, verse 1 to 8.

28. The Layman's Parallel Bible, Modern Language version, Zondervan Bible Publishers, Grand Rapids, Michigan, 12th Printing, January 1980, p. 2778. The author took the liberty of substituting the word love with the Greek word αγάπη.

30. L. Pauling, in *The Place of Value in a World of Facts*, A. Tiselius and S. Nilsson (Eds.), Wiley Interscience, New York, 1970, p. 197.

31. Pope John XXIII, April 1963, Encyclical Letter, *Pacem in Terris*. (Bulletin of Atomic Scientists **20**, 2 May 1964, p. 2).

32. Ref. 11, p. 21.

CHAPTER 2

Modern People
and
the State of Their Societies

The impact of science and science-based technology on modern man and his societies is illustrated in this chapter by the prevalence of change and violence, and by the state of man's environment and earth's resources.

2.1. CHANGE AND THE RATE OF CHANGE THAT CHANGED THE WORLD

The predominance of change in the physical world has long been recognized. Heraclitus, for example, emphasized the continuity of change. "Τα πάντα ρει" (everything is in a continuous state of flux, in a perpetual motion like a river), he said. "Change; this is what never changes," exclaimed Confucius. Nature rests by changing, poetically indicated others.[1] Like the old philosophers, the modern physicist reaffirms: nothing is at rest; science deals mostly with motion and change.

As change is with nature, so it is with man's societies. They, too, are in a continuous state of transition. Past and present societies differ, however, in the acceleration of change, that is, in the rate at which change occurs. More than in any other previous society, modern society's most distinct characteristic is the enormous change and its equally enormous rate, the prevalence of newness. This uniquely modern development has been and is being brought about overwhelmingly by science and science-based technology. Within the last fifty years alone, man has entered the atomic age, the nuclear age, the computer age, the space age, the information age, and is now stepping into the age of genetic engineering. The icons of the atom, the computer, and the gene are among modern man's most dominant symbols. An astonishing increase in man's knowledge and in man's experience, largely the outcome of the ceaseless pursue of knowledge to master and to transform nature by the man of Western civilization.

This knowledge and this exorbitant and pervasive change transforms and alters the human condition and convolutes the world: material settings, world population, urban development, accumulation of explosive power, size and complexity of social problems, interpersonal relations, values, freedoms, and responsibilities. And as it is with the given, so is with the acquired through science and technology: they can be for good or evil. Herewith, then, rises one of modern man's predicaments: he has acquired enormous new knowledge, but limited understanding; more expertise, but hardly more wisdom; greater power, but doubtfully greater personal and social responsibility. These disparities threaten to limit the societal use of scientific information and accentuate the concern that under the forces brought about by science and technology man is changing too much and too fast.

2.1.1. The New Material Settings for Life and Their Ramifications

Science and technology have shaped a materialistic society in which the modern man lives longer and better than ever before. The machinery of modern technological society feeds, clothes, cures, informs, trains, educates, and entertains him. It provides man with the benefits of electricity; the freedoms of the automobile, airplane, and telephone; the convenience of the supermarket; the services of the computer, and the Internet; the enjoyment of television; the satellite-based means of communication, information, and entertainment; the protection from disease and relief from physical pain; the enrichment of life by new man-made materials. Man has lived without most of these "essentials" for almost his entire history. Now, he possesses them and he cannot do without them.

Of the multitude of these changes and developments let us touch on only a few by way of example.

- Science and technology have caused a dramatic increase in the size, scope, and *modus operandi* of the traditional industrial and technological institutions which generate, distribute, and manage the new commodities. These industries and institutions rely on innovation and newness for success and on their scientists and engineers to keep them a step ahead of the competition. Chains of supercompanies, industrial complexes, multinational corporations, information, management, and entertainment superstructures, and military-industrial-educational supercomplexes spread a plethora of goods and services the world over (foods, drinks, cloths, pharmaceuticals, chemicals, oil, cars, machinery, communications and entertainment equipment, computers, information, services, weapons). Most of these products improve peoples' lives, others destroy them, and still others offend them or kill them; they all change them. Mass-produced goods impose an artificial uniformity on the world, and mass-delivered services and mentality wipe out diversity and tradition. All-too-powerful techno structures homogenize societies, intrude into traditional microcultural ways of life, and destabilize native institutions and governments.

- The wide-spread distribution of -- and fierce advertising for -- massed-produced goods and services has led to world-wide demands by large sections of humanity (mostly, but not exclusively, from the "underdeveloped" and "developing" countries of the world) for a fair share in these technological fruits. This demand -- "expectation" as it has been referred to -- has been strengthen by another consequence of modern science: the rapid growth in world population. The net result is that billions of persons are demanding *now* standards of living "similar" to those in the West. Never so many are impatiently seeking so much so quickly. The ramifications of such wide-spread expectations and demands for entitlement to the higher standards of living by the poor and the "left-outs" are enormous. The "revolution of rising entitlements,"[2] for which science and technology are partly responsible, is full of hope and is full of peril.

- Most of the work people do today is in fields that did not exist a century ago. They are science- and technology-based. Today, the machinery that feeds the mass production industry and the conditions under which they operate have changed the nature of the work and the worker, and the relation between the two. The standardization and replaceability of parts seems to have had a profound influence on the philosophies of the technostructures themselves, and unfortunately on the behavior of many other institutions including major scientific institutions. Standardization and replaceability has been extended to include individuals in bureaucratic organizations. The workers are standardized too; they are replaceable too. For they, too, are viewed as standard "ready-made" parts of an organizational machine for which proper training has other ready-made parts. Fitness and efficiency have imposed in many industrialized countries limits to man's freedom, philosophical and psychological perspectives, and to the basic joys of life that includes pride in a person's work.

   - Increasingly, the ability of individuals, communities, and nations to enhance the quality of life depends on their ability to access, control, process, and use information. Information is becoming one of modern society's chief raw materials and its synthesis into knowledge a distinct societal asset. Such information is principally generated by -- and provided through -- the means science has made possible: computerized services, cable technologies, global satellite communications, the Internet. These means of transmission of information along with the multitude of electromagnetic, electronic, and printed communication systems, have become enormous intellectual forces that can educate, inform, knit the world together, and unite the earth's peoples. Indeed, modern man is constantly (and instantly!) in touch with every part of the world. The strong coupling and interdependence of the peoples of the world is a new quality of the world. Through the communication systems that science and technology have made possible, old as well as new concepts of the rights of man have been and are being spread the world over and have led and are leading to demands for human rights, and the collapse of totalitarian regimes and closed-in societies. The world-wide transmission of information, the introduction of one peoples' culture into the heritage of the rest, and the diffusion of European concepts of social justice are largely responsible for the end of colonialism, the bloodless liberation of Eastern Europe, the demands for democratization of decision making, and the rapid multiplication of independent sovereign states.[3] They, especially the Internet, deprive traditional news organizations the monopoly on news reporting and the privilege to filter, manipulate, withhold, or control of the news. And this development is also a new characteristic of the world.

   At the same time these very systems become instruments of immense propaganda and powerful collectivizing forces in the hands of the bureaucratic machineries of modern governments and interest groups. The information highways spread freedoms and fresh air the world over, but they also pollute societies with unfiltered information and trash the earth. Computerized information makes it easy -- and indeed tempting -- to lie with statistics, to bias public opinion, to intrude in private life, and to limit individual and collective freedoms. Computerized societies function more and more independently of the individual, and the computer has an immense ability to reduce.[4] It seems that every step science and technology have enabled man to take has made his life more burdensome!

2.1.2. The Fragility of Society, the Tyranny of the Expert, and the Essence of Security

Modern society is too complex and too dependent on extreme detail and perfection of operation. It is a fragile society, disruptible by technological malfunctions, and vulnerable to accidents, sabotage, strikes, blackmail, or terror. It is a society which relies heavily and continually on the availability and reliability of an enormous system of systems as, for instance, the electrical supply system, the telephone system, the transportation system, the computer system, the health delivery system, and the communications system. Each and every system on which modern man depends for his daily functions, relies on the expert for its proper operation. Only the expert can operate a nuclear reactor, a high-power computer, a magnetic resonance imaging machine, or a 747 jet airplane. Only the expert can replace or repair the standard parts of the industrial machinery. Thus, to have the freedoms and the benefits modern society has delivered to him, man has to rely on and to trust the expert. This new type of almost complete dependency is viewed by many in society as a kind of tyranny, the tyranny of the expert. It has been brought about by science and technology and makes the modern man uncomfortable, if not insecure.

Modern man's discomfort with his reliance on the expert to manage his affairs is deepened by two other realizations: (i) that in a swiftly-changing world it is difficult to develop stable institutions, and (ii) that in a violent world and an era of sophisticated surveillance, atomic weapons, and intercontinental ballistic missiles, neither his private nor his national safety can be guaranteed by the state. Both have become obsolete. Both, and indeed along with so much of what man has achieved and of what man has in his possession, can disappear instantly in the flash of a moment! And this, too, i.e., the possibility of a nuclear annihilation, has been brought on him by science. Man never asked for nuclear weapons. Modern governments did at the suggestion of their scientists.[5]

2.1.3. Sweeping Social, Cultural, and Ethical Changes

Modern man lives in an atmosphere of expectancy. Constantly he wants higher and faster levels of novelty to be satisfied. Excited by the Sputnik and the first moon landing, for example, he pays little attention to the shuttle missions. His attitude becomes one of temporariness and relativism. He finds himself caught in the spiral of modern materialism, he changes attitudes and life styles, and he becomes indifferent, distrustful, uncivil, complacent. He recklessly pollutes his most precious commodity, his knowledge. He adopts -- and adapts -- to conditions that are dangerous to his physical and mental health, and unbefitting of himself. He doubts his own resilience and that of nature's. And alas! he still accepts the inevitability of war in spite of the unimaginable consequences of a nuclear war. A great deal of these attitudes and complacency can be traced to his direct contact with the many troubling, and frequently brutal world-wide events he is bombarded with daily that numb him, and to his familiarity with whole-sale unpleasantness and brutality that desensitize him. Others can be traced to the unlawful behavior of the nation states and the disrespect shown by their governments for international and human law (Section 2.2).

The hurricane of change shakes the modern state and diminishes its authority. The relentless undermining and destruction of traditional institutions and the ideology, values, and ways of life they represent, strike the self at its core and breed insecurity. Invading technologies and competing ideologies fragment man's smallest institution, the family. The ferocious winds of change erode man's purposefulness and ethical sensitivity, depersonalize, and bury the timeless in the debris. The moving music of the Gypsy, the unselfish generosity of Santa Claus, and the free-spirited dance of Zorba gave way to synthetic happiness. Even the peculiar honesty of the Athenian of 25 centuries ago -- the one who when asked why he voted to send the honest Aristides to exile replied that he did it because he could not stand his honesty -- even that peculiar honesty has become a rare human quality. Precious human attributes that have graced the face of man for millennia are all being pushed aside to make room for change.

At the foundation of man's ethics, it has been correctly said, is not "the knowledge of the scientist or the expert, but knowledge of a kind readily available to all men of good will."[6] And yet man's ethics need stretching beyond their former terms, farther than their traditional horizons of distance and time. Science forces a new role of knowledge in morality. Thus, the ceaseless change must allow time to develop an ethic for itself. An ethic for change that would keep things in perspective and would interlock the technological and the social systems. An ethic for change that would enhance modern society's confidence in science and in itself, for modern man is fearful of science and of himself. He does not trust himself to handle responsibly the enormous power that has been handed to him by modern science. Modern science has made man desperately afraid.

Life has a way of coping with change if it is given a chance, and so can man. The threat that change has bred, said anthropologist Margaret Mead, can only be solved by change itself. Such a change will continue to come from science, but will it also come from within man himself?

2.2. A VIOLENT WORLD WITHOUT PEACE

If change is the most distinct characteristic of modern times, violence follows closely. Ours is a distinctly violent world, a world without peace. In the post World War II "peace time," well over 100 major wars and conflicts raged, and some are still raging, the world over: in South America, Central Europe, the Middle East, the Persian Gulf, in Central and South East Asia, and in Africa. Each of these wars has been a source of suffering and destruction. Each became a river of sorrow and fear. Each has displaced and has depreciated yet more humanity.

Along with these wars and conflicts we have witnessed the rise and the fall of the Berlin wall; the strengthening and the collapsing of the "Iron Curtain;" the massacres in Vietnam, Cambodia, Biaphra, Rwanda, and Burundi; the bombings in Oklahoma City, the gassings in Tokyo, the brutalities in Tiananmen Square; the ethnic cleansing in Turkey, Bosnia, and Kosovo; blackmails, skyjackings, murders, antisocial outrages of groups and individuals against man and society; indiscriminate violence and boundless crime the world over. Ours is an era of national and international anarchy, disrespect for national and international law, abuse and violation of civil liberties, crimes against humanity.

This high level of antisocial activity is blamed on the weakening of traditional values, but also -- especially its effectiveness -- on the world-wide spread of *technical* means of destruction which have been made available to man through science and technology. New technologies of violence are upon us that enable individuals and tiny groups to easily wield unprecedented destructive power. There is, unfortunately, another important factor which contributes to the disrespect of the citizen for the law: the unlawful behavior of the governments of nation states. Nothing can illustrate this last point more graphically than the reckless behavior of such states regarding *the armaments and the method of war.*

Today, the governments of many nations possess armaments of cataclysmic proportions and consequences. Let us just focus narrowly on nuclear weapons. Until recently, five countries [the United States of America (USA), the Former Soviet Union (FSU), the United Kingdom (UK), France, and China] possessed nuclear weapons. Two other countries, India and Pakistan, conducted their first nuclear tests in May of 1998. At least two other countries, Israel and South Africa, have developed nuclear weapons or have the capability to rapidly assemble them, and still a number of others have secret plants to produce fissile materials (separated plutonium or highly enriched uranium).[7-12] Since the explosion of the first atomic bomb on July 16, 1945, (Figure 2.1), the testing and stockpiling of nuclear weapons has been beyond reason. Numbers for the latter for the two major nuclear powers from 1945 to 1994 (and also for UK, France, and China) are shown in Figure 2.2 and clearly exemplify man's nuclear predicament. Although the nuclear arsenals of both the USA and the FSU are declining, a destructive capacity of immense magnitude is still in the possession of these two countries. In 1994, the USA alone, had an estimated stockpile of 14,900 warheads[14] corresponding to a total destructive capability of 2,375 megatons (MT) which is equivalent to about 200,000 Hiroshimas.[15] Potentially this arsenal (and a similar one in the FSU) is capable of killing everyone on earth many times over. This destructive power peaked in 1960 when the USA had 20,491 MT capacity.

Today's high-yield fusion weapons are much more powerful than their fission predecessors. A single bomb's destructive capacity can be as large as nine million tons of TNT[14]. Science and technology have increased not just the warfare megatonnage, but also the warfare's capacity for precision. It has made the warfare technology fearfully efficient!

Let us sharpen our focus a little more on *nuclear weapons and nuclear war* and address a few questions to ourselves. First, why do we have

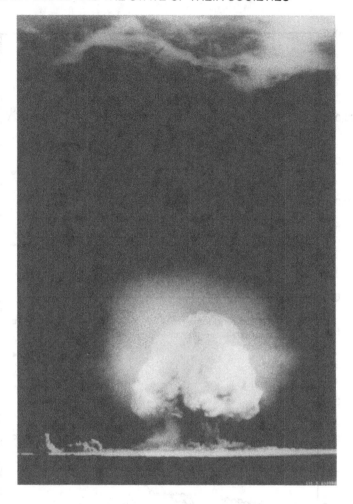

Figure 2.1. The first atomic bomb explosion, the "Trinity Test," on July 16, 1945, a fortieth of a second after the start of the explosion.[13] It took place at Jornado del Muerto, near Alamogordo, New Mexico (Photo LANL CIC-9, provided by Los Alamos National Laboratory).

nuclear weapons? The answer, I suppose, must be because science and technology provided the knowledge and the capability to build them, and because society, or its political leaders, considered them necessary for security. Second, why do we feel that we need to have so many nuclear weapons, many more than we can rationally justify, to defend ourselves?

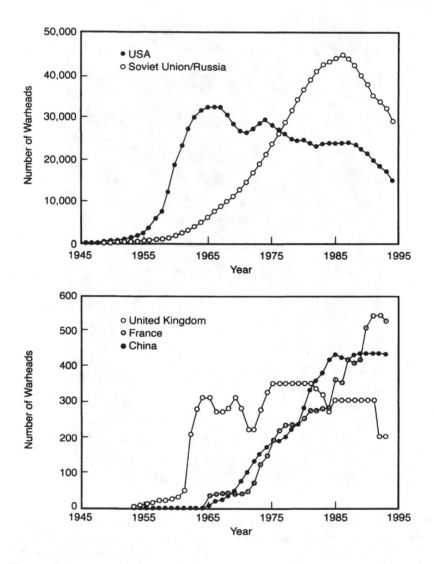

Figure 2.2. Total stockpile of nuclear weapons by the United States and the Former Soviet Union from 1945 to 1994 (top) and by UK, France, and China (bottom) (based on information given by Norris[14]).

The answer, I suppose, must be because we *fear* that our adversaries may strike first and annihilate our ability to retaliate; we want to be sure of their mutual destruction! Third, why do we keep investing at such exorbitant levels in more and more sophisticated arsenals in spite of our other pressing needs? Because, I suppose, we *fear* that the other side (or sides) will get ahead of us. Because the *frontiers of science and technology have truly become the frontiers of weaponry*. Because scientific inquiry and technological innovation cannot be stopped and consequently the only way to stay ahead, as a United States Defense Department official put it, is for our scientists to get smarter and our engineers to be able to make the technology for the next level of warfare technology. And so, under the constant presence of the fear of war, we have developed, through our government's support of science and technology, atomic bombs, hydrogen bombs, neutron bombs, powerful and accurate delivery systems equipped with some of the most sophisticated computerized guiding devices the world has ever known, and we are still developing speed-of-light weapons for use on earth and in space. And all this we do at a staggering cost: economic, environmental, intellectual, mental, emotional, spiritual, moral.

In the name of security and defense it is argued. And yet, as we already indicated, our security is evasive and our defenses are vulnerable. No nation can win a nuclear war. Science and technology can offer the world no real defense against the consequences of nuclear war. Science clearly speaks of only the horror (the deadly damage by explosion, radiation, and heat).[16] The invention and development of nuclear weapons represents a qualitative break in the history of violence,[17] for it is now possible to destroy human life on earth.

In the name of peace, *peace-through-deterrence*, is maintained by others. The doctrine of "peace-through-deterrence," it is argued, has prevented a nuclear conflict[18] and has forced containment of the many small wars that tarnished the face of the earth since WWII. But can the world continue to pretend that the nuclear weapons and advanced armaments we have been asked to produce in "peace time" are the guardians of peace? Should we continue to accept the thesis, as many have asked, that because since the destruction of Hiroshima and Nagasaki another atomic bomb did not go off proves that the miracle of fear called "peace-through-deterrence" works? Should we, in effect, accept this view, implicit in the statements of some of the scientists who built nuclear bombs, and "love the bomb"?[19] How can armaments be the *permanent* arbitrators of peace? If we rely on terror to keep the peace, is it not obvious that "the worse the terror, the better the chance for peace?"[20] And is this not, a diverging process which will ultimately lead to the inevitable explosion? History is as cruel on this issue as is Murphy's law (anything that can happen will happen, and anything that can go wrong will go wrong): *it is a fact of history that every means of destruction man has ever invented -- including the atomic bomb -- has been used in war.* Should we, then, follow the Delphic Oracle's example -- which broke only once its traditional vagueness and predicted, in its last oracle (χρησμός), with full certainty its own fall -- and predict with certainty society's holocaust? How can humanity be expected to live for ever under the threat of nuclear annihilation? We *must* resolve that nuclear weapons never be used, never again. And we *must* further resolve that the ethic of the arms race is replaced by one of arms control and elimination. This is a fundamental necessity of man to which he and his governments, however arrogant and however distrustful of each other, must bow. Science and technology can help man trace, control, store, and destroy these weapons. Better yet, science and technology can help man use their deadly material to generate electricity. A skeptical citizen has every reason to distrust the expert and the politician alike on this issue.

And yet, there is hope. The expert and the politician have taken fundamental and hopeful steps:[21,22] The Partial Test Ban Treaty initiated by President Kennedy in 1963; the Biological Weapon Convention initiated by President Nixon and signed by President Ford in 1975; the Strategic Arms Limitations Agreements between the USA and USSR and Russia; and the Disarmament Agreements between the two superpowers (adversarial nation states agreed to mutually reduce their arsenals.); the reasonable success in limiting the spread of nuclear weapons.

In spite of this progress, the nuclear threat is still upon us undiminished. Many nation states are determined to acquire or to strengthen their nuclear capabilities. Some, like France and China, were testing until 1995, and India and Pakistan joined the nuclear club in earnest in May of 1998. All nuclear powers, but especially the United States and the Former Soviet Union, have enormous nuclear contamination problems from weapons production complexes.[23] The dismantling of the Former Soviet Union has presented new threats regarding the safeguarding of nuclear weapons and materials.[24] And alas! the developmental evolution of weapons and warfare methods and the stockpiling of bombs-- conventional, atomic, thermonuclear, laser, neutron, chemical, bacteriological -- by the institutions of war continues (or can be activated at any moment) hand-in-hand with their most indispensable partner: science and technology. A number of scientists -- most recently Hans Bethe in a letter[25] to the President of the United States-- urged to halt nuclear weapons research and to emphasize instead their stewardship.

The competition between the major world powers in production and refinement of arms is matched by the fierce selling of the means of destruction. Big and small nations are arm merchants. They sell promiscuously big, bigger, better, and more expensive arms! Arms sales is a major source of income (and political influence) for the major powers, and the sustaining force of the conflicts in the many uneasy areas of "third-world" countries, the very countries where the majority of the people are hungry and desperate; the regions of the world where most of the post WWII conflicts have occurred and are still occurring. And, as if to complete the process, the superpowers are not just arming, they or their surrogates are also training the armies of the smaller countries.

At the same time, science and technology made it possible to instantly globalize war and human conflict. In doing so, they made it possible to deprive governments of their capability to make war in secrecy. These are hopeful developments for they constitute an effective *nondestructive deterrence* of war.

War making and violence make modern man fear the world he has created. Unsecured, he is detoured from positive action, wastes resources, and buries the gift of providence. He loses the spirit of hope he so desperately needs to curb violence and war. Man needs to understand better the causes of war and with the aid of science to diffuse them.

2.3. MAN'S NATURAL ENVIRONMENT

At all places and at all times man has been hard at work to change the natural environment to his advantage. Most of humanity, for instance, has lived on "man-made" lands created since the agricultural revolution. Man's ability to metamorphose the surface of the earth is miraculous. He has brought to light the richness of nature, and beautified and shaped the face of the earth according to his own culture, to paraphrase Dubois.[26] Over the centuries man created patiently and carefully parts of the earth "in his own image." He cared about the earth and the life it sustains. With the aid of science and technology he managed admirably parts of the earth as a responsible custodian.

At the same time, the human environment has dramatically and irreversibly changed as a result of scientific and technological developments. For instance:

- Ten thousand years ago there were no metals in the human environment (metals are rarely found in the pure form in the environment); we now have a metallic environment.[27]

- Besides lightning and frictional electrostatics, electricity appears rarely in nature in observable forms; we now have a completely new electric environment.

- Apart from the rare cases of natural radioactivity, nuclear phenomena on earth are man-made; we now have conditions (atomic bombs, nuclear wastes, nuclear reactors, artificial radioactivity) for a new "radioactive" environment that can stress the fabric of life. Man has lived in sensitive equilibrium with his radiation environment up until the end of the 19th century.

- Most of the chemical substances we now use directly or indirectly did not exist 100 years ago; we now live in a new environment of man-made chemicals growing at a rate of many thousands each week.

The face of the earth has been changed by these and similar other new environmental conditions. It is no longer possible to dissociate progress from its consequences. Man continues to create better conditions for life, to beautify, and to groom the face of the earth, and, simultaneously, he continues to destroy, interfere with, and overtax nature. He continues to engage himself in a devastating war against his own home: polluting, contaminating, wanting to touch and to change everything, often, it seems, without reason. A psychology prevails that justifies any kind of change if it is economically profitable.

Harmful synthetic chemicals and dangerous by-products of combustion pollute the air we breath, pesticides contaminate the food we eat, industrial chemicals and nitrates from agricultural fertilizers pollute our streams and lakes, and radioactive strontium from nuclear explosives finds its way to our children's bones.[28] The burning of fuel by internal combustion engines, themselves valuable sources of energy and personal freedom, creates air pollution, acid rain, and generates ozone depleting gases (e.g., sulfur and nitrogen oxides)[29] and global warming gases (e.g., carbon dioxide).[30] While the extent to which man-produced greenhouse gases have caused an increase in the earth's temperature is still debated, the increase in the amount of carbon dioxide in the atmosphere is unquestionable as can be seen from Figure 2.3. Drugs widely prescribed for their health benefits, are found to have harmful side effects, as for example, thalidomide which was used in the 1960s as a tranquillizer and later found to be associated with human foetus malformations, or DDT, which is still the cheapest pesticide

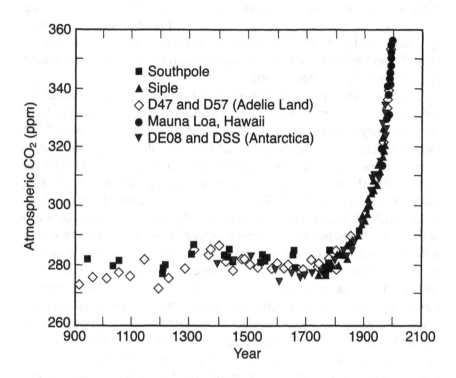

Figure 2.3. The amount of CO_2 in the atmosphere has increased rapidly from its pre-industrial level of about 280 parts per million (ppm) to about 360 ppm today (from Joos[31]).

to use against malaria,[32] but contaminates the land, the food and the water and accumulates in our bodies. Nuclear reactors, a new source of energy, generate dangerous radioactive debris and carry with them the possibility of catastrophe as the Chernobyl experience can attest.

Whatever new material man places in, and whatever new technology man introduces into the environment, becomes part of the environment and plays a role thereafter. Useful new substances and beneficial technological innovations often turn out to be problematic and harmful in ways unforeseen. There is a risk associated with them along with the benefit. Indeed, man has learned to accept the risk in order to reap the benefit. How many of us, for instance, would refuse to accept the risk of having electricity in our homes or to drive our cars? And yet the risk of been killed by either is real. It is through science and technology that we can identify, quantify, and minimize the risk. And it is through science and technology that we can maximize the benefit.

And yet, at any given time, science and technology may be unable to give clear-cut answers to open questions because at the time the questions are posed they are *trans-scientific*.[33] They can be scientifically defined, but they cannot be scientifically answered (Section 10.3). Many such issues arise when one is attempting to identify and quantify possible health (or biological) effects of low level insults let them be chemicals, ionizing radiation, or magnetic fields. The increased significance of possible low-level health effects of otherwise benign and beneficial technologies can perhaps be illustrated by the current debate on possible adverse health effects -- including cancer -- of long-range exposure to low-frequency magnetic fields associated with the countless ways in which electricity is woven into the fabric of modern society (via lights, appliances, motors, power transmission and distribution lines, microwave equipment, and the myriad of ways we consume electric power). Field exposures are an inevitable consequence of living in a society that uses electricity. And when society uses electricity in as pervasive a manner as the present society does, science is called upon to provide an answer even though at present any possible health or biological effects seem to be buried in the noise. Although some epidemiological studies claimed to have observed a positive association between magnetic field exposure in home environments and cancer, particularly in children, this claim has neither been supported by other studies nor has it been accepted.[34] It is desirable to seek answers to the possible effects of low-level insults by searching for the basic mechanisms via which these insults interact with the living system or the cells. At the mechanism level, any possible effects can be detected early and their harm avoided or ameliorated. This is a role of, and benefit from, basic science.

Let us complete this discussion by noting a few elements of man's recent concern about his environmental interventions. For man has become increasingly and worrisomely aware of the state of the natural environment and many factors contributed to this realization.

- *The deepening awareness of the preciousness and uniqueness of the earth as the human home.* The human eye has seen the earth from the darkness of space as a naked little star (Figure 1.1), the only star man can turn to. A naked little star, but one with air and water and resources in its thin envelope to sustain life. A precious home, inherited and created. Like the giant Antaeus of the Greek mythology man retains his strength when in direct contact with the earth. He is bound to the earth, wedded to its crust. Unfortunately, more and more of us are losing intimate contact with the soil. Most of us never marvel, for instance, at the mysterious power of the earth's soil to give food that makes life possible. I grew up in a big city, a colleague once told me, where all I saw was asphalt and concrete, and soil was for me the dirty stuff they dig up when they tear the street to fix something.

- *The "shrinking" earth.* The earth is no longer big and endless. It never was. Dubois estimates that of the earth's total land surface 11% is used as cropland, 10% as range land, and 20% as managed forest. The largest part of the remaining 59% "is almost constantly frozen or is too cold or too mountainous for normal human occupation or utilization."[35] As population, consumption of energy and resources, housing, communication networks, highways, and the standard of living of the late 20th century industrial and post-industrial societies increase, so do the demands for land. Man is running out of the earth's land surface compatible with human life.

- *The fragile web and the mutual dependence of life.* There is a delicate balance between the earth and the life it sustains. Nature's cycles are mighty, but delicate, and the restoration capacity of the earth is finite and rate-limited. Most of nature's states are finely tuned dynamic equilibria. Small deviations from these equilibria caused by man disturb this delicate balance and strain the earth's resilience. Modern man has become increasingly aware that he has to live in harmony with nature, cooperate with it, and adhere to the golden principle of minimum intervention. He is becoming increasingly cognizant of the mutual dependence of all that exists on earth; that on earth everything is relevant to everything else, and that he, man, has to respect the holiness of the whole and the sanctity of the part, the needs of each particular area and of each individual species. Man increasingly realizes that he alone can willfully damage the web of life on earth. This realization makes him increasingly concerned of what he destroys to accommodate his modern ways of life and habits. Through science and technology he can fulfill his complementary roles as master and as servant/custodian of nature.

- *Interdependency of human functions.* The size, severity, and implications of the many environmental problems and complexity of environmental issues, led man to a deeper realization of the interdependency of human actions and responsibilities. Most of the environmental problems are neither local nor regional, they are common with one's neighbors, and one's neighbors is everyone on earth. They know no boundaries and relate to all forms of life and their interrelations. How, then, can any nation claim as its sovereign right to pollute the air and the water of its neighbor states, or to spread over their fields the radioactivity of its nuclear works?

- *The quality of life.* Man is worried about the destruction of the biosphere and its suitability for life. He is worrying about the imbalance between his organizational systems and the earth's capacity to sustain life. As the environmental conditions on earth change, so do the living things on earth; they have to adapt. To do so they have to evolve. And herewith lies another of modern man's fears: *lest the exceptional capacity of man to adapt leads him to accept environmental conditions increasingly unbefitting of himself.* Fear for a degradation of the quality of life. Adaptability, says Dubois, is "an asset for biological survival, but paradoxically, the greatest threat to the quality of human life is that the human species is so immensely adaptable that it can survive even under most objectionable conditions."[36]

- *The concern for posterity.* Man has come to realize that he is imperiling future generations by shifting the cost of his actions to them. Western culture's "dominion and multiplication" is disastrous if it is not accompanied by a responsible management of the earth. The Judeo-Christian tradition speaks of "dominion and multiplication," but it also speaks of stewardship, responsibility, and accountability.[37] Man is, thus, instructed -- as his ancestors had been long time ago -- to share the earth, to revere nature, and to accept responsible stewardship of the earth. And a responsible stewardship of the earth calls for the marshaling of the forces of science and technology on behalf of the earth and the life it sustains: to control pollution, conserve resources, reduce waste, recycle materials, replace "needed" but harmful chemicals, discover and develop new environmentally friendly materials and energy sources; set environmental standards and regulations, assess the risk and the benefit of new scientific and technological developments, and establish a basis for ecological ethics; learn about the environment, clean the environment, and repair the damage that has been inflicted upon it by man. It is for instance disturbing that most of the contaminants of the urban air are still unidentified, and their sources of origin, reactions and pathways in air, water and land and biological effects are still largely unknown. Moreover, since man's natural environment will never return to its pristine preindustrial condition, there will always be residual amounts of diverse physical insults in the biosphere. Science and technology will enable a responsible steward to detect them, identify them, and find ways to cope with the unavoidable problems of *low-level* physical and chemical insults of the biosphere and the *subtle* environmental factors in general. A responsible stewardship of the earth is personal and collective, and can be made possible with the knowledge and understanding that science provides and the tools that technology invents.

2.4. EARTH'S RESOURCES

The earth is rich and resourceful, but each new year man requires for his support more food, more raw materials, and more energy. Not only proportionately more people consume more than the year before, but each new year there are more people the earth must support. It took all of history to about 1850 A. D. for the earth's population to increase to one billion, but it took less than 100 years to reach 2 billion by the 1930s. Thereafter, one billion people were added to world population every 15 or so years, reaching 4 billion by the year 1975. World population is now heading toward 6 billion and United Nations projections indicate[38] it would increase to somewhere between 10 and 16 billion before leveling off. The proper consideration of the resources of the earth, therefore, is intimately linked to world population, to the standard of living of the peoples' of the world, and to the unavoidable resultant impact of both on the environment. It is also strongly coupled to the way the resources of the earth are managed through science and technology. The understanding, management, effective use, preservation, and expansion of the resources of the earth have two irreducible components: the sociopolitical and the scientific/technological. To illustrate the role of the latter let us narrowly focus on three basic commodities: food, natural materials, and energy.

2.4.1. Food

From agriculture, farming, and food production, to new foods, food enrichments, food supplements, food packaging, and food distribution, modern man relies on the methods of science and on the tools of science-based technology. The availability of energy, machinery, and the advancements in chemistry revolutionized man's food supplies. This revolution is now joined by biotechnology and genetic engineering.

Biochemical research in the 19th century led to the invention of inorganic fertilizers, by Justus Liebig, a scientific contribution that averted the first nutritional catastrophe of modern times.[39] Scientific discovery has averted then the perceived imbalance between food availability and increasing food demand due to population growth, and has made since Malthusian pessimistic predictions concerning humanity's future *potentially* obsolete. Chemistry in particular has been an essential element in achieving the remarkable increases in agricultural productivity, especially during the last half of the 20th century.[40] For instance, from 1950 to 1980 world food production doubled, with the increase in the developing countries exceeding that of the developed countries,[41] principally due to the development and adoption of fertilizer-responsive cereal varieties, such as wheat and rice, and the use of pesticides.[42] Today's chemistry is an integral part of agriculture and food processing technologies, and a determining factor in the amount and variety of foods available for human consumption. Modern man's science and technology have made not just the Malthusian ideas, but also Darwinism, *potentially* inapplicable to modern man: man's total hereditary factors are no longer a result of the struggle for survival.

Increasingly, biotechnology and genetic engineering complement conventional chemistry with new opportunities and approaches to agriculture. For instance, recent advances in gene manipulation have introduced a new dimension to the future of crop production. Through gene combinations it might be possible to create plants that would not need fertilizers.[43] The scientific/technological challenges and opportunities in food supplies that lie ahead are thus enormous. They spread from the development of tropical agricultures to new ways of farming the land and the sea, from new machinery technologies to new sources of energy, from chemistry to biology and botany, from a limited[44] to a more diversified food supply. Energy use continues to shrink the land needed for food production. Science and technology make it possible for man to feed himself. Regrettably, he has not as yet mastered the will and the sociopolitical skill to do so for all peoples.

2.4.2. Natural Materials

Besides food, man needs many other types of materials for his life. Most of these are free and inexhaustible, others are not. Besides air (oxygen and nitrogen) and water, man needs iron, aluminum, copper, zinc, tin, lead, and so on. For his fertilizers he needs phosphorus and potassium. At the beginning of this century only about 30 chemical elements were in commercial use, there are more than 70 in use today.[45] The highly industrialized modern society steadily needs more quantities of these and other materials such as fossil fuels. Albeit few notable exceptions such as crude oil and natural gas, the earth is rich in mineral supplies[46] provided man has availability of energy and effective ingenuity through science.

The advancement of science - - and especially the progress of technology - - have depended on the discovery of materials with new properties (mechanical, chemical, or electromagnetic), and on the invention of techniques to process them. Historians identified distinct steps in man's civilization by his ability to deploy minerals: the stone, the bronze, the iron, the coal, and the nuclear ages. New developments in the treatment of metals, ceramics, fibers, glasses, cellulose and rubber derivatives, and composites generated new technologies and new industries. And so did the invention of an economical filament for the electric light bulb, the invention of the transistor, the invention of the laser, and the development of the light-weight aircraft engine.

2.4.3. Energy

It is clear from the preceding discussion on food and natural materials that they are both tightly tied to the availability of energy. More than any other resource, energy has shaped man's past and will shape man's future. Man's muscle gave way to tools and machines: the axe, the plough, the wheel, the steam engine, and the steam, hydroelectric, and nuclear power plants. His cultural and social development was made possible by his unique ability to exploit energy sources outside his own body and beyond the wild food and the ambient heat of his environment.[47] *Man never refused to develop an energy option presented to him.* All industries, wrote Maddox,[48] are devices for turning materials of one kind into some other form, and all of them require energy to be effective. Energy is the ultimate raw material. It is modern man's slave. Energy can be transformed from one kind to another, but it cannot be recycled.

The energy resources of the earth trace to the sun. The sun bathes the earth in a huge amount of energy. It had done so for millions of years. To quote Cook,[47]

> "Life exists on earth by the grace of solar radiation. It provides the ambient temperature range within which organisms can survive, drives the winds, and powers the hydrologic cycle. It provides the energy for plant growth and the light without which there would be perpetual night. Without plants there would be no animals, no humans. Over millions of years a tiny part of the incoming solar radiation has been trapped in the form of plant and animal detritus, preserved, and turned into deposits of petroleum, coal, and natural gas -- the fossil fuels upon which modern industrial man is utterly dependent."

Modern man not only learned how to better use the energy provided by the sun in terms of fossil fuels and in terms of direct every-day radiation, but he also learned through science the secret of the origins of sun's energy, kindled billions of years ago. The modern scientist has been likened to the Prometheus of the Greek mythology in that like Prometheus he, too, has learned the secret of fire from the gods and passed it to man. It is now, as others have already pointed out, up to man himself to use this knowledge to develop inexhaustible supplies of energy and thus abundance of food and materials, or to burn himself to death and blame science for it!

Past, present, and future energy sources are intricately connected with two special branches of science: thermodynamics and nuclear physics. Thermodynamics is a relatively old field of science governed by some of the most general principles known to science (Chapter 3). It rests on two fundamental laws: the first law of thermodynamics -- the law of energy -- and the second law of thermodynamics -- the law of entropy. Both are of universal validity: they apply to all forms of matter and energy and their exchanges under all conditions.

The first law of thermodynamics tells us that the energy of the universe is conserved. Or, more appropriately for our discussion, it tells us that a system can do work when heat is added to it. For a closed system (i.e., one whose mass is constant) to which an amount of heat Q has been added, the work W done by it equals the amount Q of heat supplied less the change ΔQ in the internal energy of the closed system, i.e.,

$$W = Q - \Delta Q. \tag{1}$$

This simple law was not formulated until the 19th century because it depended on the interpretation of heat as a form of energy. It is an extraordinarily simple statement of profound significance for both science and society. Prior to its establishment, man relied on his muscle, the muscle of the beast, and on the flow of the water or the blowing of the wind as his sources of work: man and the beast, thus, became commodities to be used for this purpose. Man enslaved man to do work. The first law of thermodynamics, the simple equation above, has changed all that: *work can be done from heat*. There is no longer a need for slaves! Construct an engine that can change heat into work and free man from slavery![49] And so, the first law of thermodynamics led to the development of the heat engine -- the first example of which was the steam engine -- a landmark that signified the beginning of the industrial revolution.

The second law of thermodynamics deals with entropy. Unlike energy, entropy is not conserved, it always increases. Entropy is a measure of the unavailability of a system to do work. An increase in entropy is accompanied by a decrease in energy availability. Hence, the second law of thermodynamics tells us that what eventually diminishes is not the world's stock of energy, but its ability to do work.[50] The first law of thermodynamics says that it is possible to use heat to do work and the second law of thermodynamics prescribes how to do it efficiently. Based on it, as the work of Sadi Carnot has first shown (Carnot cycle), one can determine the efficiencies of heat engines.

While still man's energy production relies heavily on thermodynamics, the science of energy found a new home: modern physics (nuclear physics in particular). That this is so, can be traced to another elegantly simple equation, Einstein's mass (m)/energy (E) equivalence, established at the dawn of modern physics at the beginning of the 20th century,

$$E = m c^2. \tag{2}$$

Or, for our purpose,

$$\Delta E = \Delta m \, c^2, \tag{3}$$

where c is the speed of light in vacuum. According to Equation (3), if a small amount of matter of mass Δm can be converted into energy ΔE, large amounts of energy are released. Such an energy source can be available to man to use. Modern physics has shown this to be feasible in two fundamental, "primordial" ways: *nuclear fission and nuclear fusion.* Physics has told man how to fission heavy elements (the fissile elements of uranium, plutonium, and thorium). The first controlled nuclear chain reaction was achieved at the University of Chicago on December 2, 1942, in the midst of WWII, and nuclear power from controlled nuclear fission became a reality a few years later. The burning of uranium does not involve oxygen and is a very efficient source of energy. One ton of uranium yields the energy equivalent of three million tons of crude oil. Physics, also, told us how to fuse light elements (e.g., deuterium and tritium) and obtain potentially even larger amounts of energy from flameless burning of matter. Nuclear power has not as yet become available from controlled nuclear fusion, but this is just a matter of time. Uncontrolled power in the form of atomic bombs has, of course, become available to man from both nuclear fission (in 1945) and nuclear fusion (in 1952). Nuclear power came from basic science and in this regard it is a purely scientific form of power.

The main sources of energy available to man today fall into three broad categories: fossil, renewables, and nuclear.

2.4.3.1. Fossil

The main fossil fuels are natural gas, oil, and coal. The most abundant of these is coal. Coal represents about 90% of all known conventional fossil resources. There are differing views as to the estimated size of fossil fuels, and a spectrum of opinion on how or whether to burn such valuable resources. However, independently of these differences, there seems to be wide-spread concern that natural gas and oil will be exhausted within a few centuries and that coal will not be able to meet man's expanding energy needs. This becomes more of a problem if one considers the consequences of using huge quantities of coal: the adverse effects of sulphates, nitrous oxides, carbon dioxide, particulates and aerosols, heat, heavy metals (such as mercury and cadmium), radioactive materials (such as radon) in coal fly ash, and the huge quantities of debris. Neither the health nor the climatic effects of these factors have been fully assessed, but they are thought to be serious.[51] The increase in CO_2 in the atmosphere from fossil fuel combustion has been demonstrated (Figure 2.3) and is most worrisome because atmospheric CO_2 helps regulate the temperature profile of the earth's atmosphere and surface.

2.4.3.2. Renewables

The category of renewable energy sources comprises hydro, solar, biomass, wind, and geothermal. Today, of these sources only hydro (generation of electricity by the flowing of water) is a significant source of energy contributing about 20% of global electricity (compared to about 17% of global electricity generated by nuclear).[46] Compared to coal, hydro has no fuel cost, adds no extra heat to the environment and has a low maintenance burden. Hydro's expansion will, however, be limited, mostly for environmental reasons: it involves flooding of large areas and altering the flow of rivers.

The energy of the sun provides man a multitude of energy sources. Direct solar radiation and indirect solar-radiation-based energy sources such as biomass, wind, and geothermal are all due to the energy received by the earth from the sun -- the nuclear fusion reactor located a distance of some 150 million kilometers from the earth. Solar radiation as an energy source is inexhaustible, and it is free. It is, however, diffused and intermittent and must, thus, be trapped and be looked at in combination with bioconversion and energy storage. Solar radiation is used in basic conversion processes utilizing solar heat or light. Solar technologies, then, generally fall into three classes: thermal, biomass, and electricity generation. Solar thermal technologies are usually used for heating and cooling buildings. Solar radiation arriving in the form of light is most prominently utilized by nature in photosynthesis (a green plant remains the best solar energy capturing device!). Historically, this has been the energy source of man in the underdeveloped regions of the world: wood, shrubs, agricultural wastes, and animal dung. Less than 2% of the global energy supply is estimated to be from such sources.[46] Recent scientific efforts have been directed[52] to finding plants from which liquid fuel can be extracted and used directly as an energy source (e.g., sugarcane to produce alcohol for use as a fuel additive). Solar radiation can be also converted directly to electricity by means of photoelectric cells. Much scientific effort is devoted to photovoltaics, a high-technology approach to converting sunlight directly into electrical energy. A photovoltaic device in its simplest form, is a solar-powered battery whose only consumable is the sunlight that fuels it. Single-junction solar cells are made from crystalline silicon and GaAs and have high efficiencies (about 25%).[53] Photovoltaics are environmentally benign sources of electrical energy, but they are still not competitive cost-wise with conventional forms of electricity.

Another source of energy, geothermal, results from the heat in the earth that is available near the surface of the earth's crust and is used mainly for electricity production and direct heat applications. Geothermal energy sources are not expected to contribute significantly to the needs of modern industrial society.

2.4.3.3. Electricity

It seems safe to say that society is perennially hungry for energy. Man never has all the energy he needs. But today's society, more than any other in the past, is particularly hungry for a particular form of energy: *electricity*. We single out electricity not because it is a source of energy (the other sources of energy discussed here convert to electricity), but because electricity is the fluid which sustains modern man's technological society. Electricity's versatility, controllability, cleanliness, end-use efficiency, and ease of use have made it a continuously growing choice for efficient economic systems even though the cost of producing electricity makes it the most expensive form of energy per unit of energy content. The extraordinary range of new applications of this remarkable form of energy includes microwave ovens, industrial lasers, opening and closing computer circuits, electronic process controls and an endless list of potential new uses. In the USA, electricity production by the utility industry[54] between 1948 to 1978 has increased almost eightfold, while the total energy consumption rose by only a factor of 2.4. Today, in the USA about 40% of all primary energy equivalent goes into electricity generation.[46] Electricity's historic limitation has been the low energy density of electrical storage devices, principally the battery, as compared to the energy density of petroleum products and coal. Improvements in battery technology and fuel cells are crucial for such new applications as the electric car and airplane. Electricity's ultimate grain is the electron, and electricity's production, distribution, and use are based on fundamental science. Electricity can be produced by nuclear power.

2.4.3.4. Nuclear Power

Nuclear power is of two types: *fission* and *fusion*. Nuclear power from fission reactors is a reality. Nuclear power from a fusion reactor has not as yet been achieved. It remains a scientific and technological challenge. That the power of nuclear fission has been harnessed for the benefit of mankind is a remarkable scientific and technological accomplishment. In the nuclear fission process *heavy* nuclei are broken up into fission-fragment nuclei by the absorption of neutrons, minute particles that enter the positively charged nuclei of the fissile material easily because they carry no electric charge. These neutrons are themselves produced from neutron-induced fissions in a chain reaction involving the uranium isotope $^{92}U_{235}$. Nuclear fission is presently utilized in thermal reactors, whereby some of the energy released in nuclear fission reactions is used to generate heat which in turn generates electricity. In the USA the most common type of thermal reactors is the light water reactor which uses uranium slightly enriched in the isotope $^{92}U_{235}$.

There is, of course, another possibility of achieving a long-range source of energy: the breeder reactor. Here $^{94}Pu_{239}$ and $^{92}U_{233}$ are used as the catalysts for burning $^{92}U_{238}$ and $^{90}Th_{232}$ which are in abundant supply. The breeder reactor offers a long-term source of nuclear power, but besides the problems it shares with conventional nuclear reactors in terms of the acceptability, it presents the security issues of a plutonium-based fuel cycle and requires a more elaborate cooling medium (molten sodium). The latter requirement is because, unlike conventional reactors the breeder operates on *fast* neutrons so that a neutron slowing-down medium such as water cannot be used. Instead, in order to transfer the heat that they produce, the fuel rods are immersed in molten sodium which is pumped through a heat-transfer system that eventually generates the steam that drives the turbines.[55,56] Regrettably, there is no breeder-reactor program in the USA today. France, the only country in the world which developed the breeder reactor, has recently halted that effort.

Nuclear power from fission does not as yet have broad public acceptability and continues to have serious obstacles (mainly sociopolitical). It appears that technically the major problems of safety and waste disposal are largely solved or solvable, but solutions are being delayed for sociopolitical reasons. Society is concerned about reactor safety and accidents. Nuclear accidents are low probability events. Nonetheless, operational safety breaks down and accidents do happen, as those at Three Mile Island (1979) and Chernobyl (1986). In particular, the Chernobyl catastrophe brought home the realization that "a nuclear accident anywhere is a nuclear accident everywhere."[57] If man is to live with fission over the long term, it is argued,[57] "he must reduce the *a priori* probability of accident by a large factor, say 100." Others argue for a larger factor.

Society ties nuclear power to radioactivity and its accumulation and intrusion into the environment, although the production of low-level radioactive waste is not confined to the nuclear-power industry. Large quantities of low-level waste, for instance, are produced by the medical establishments. Here is how Novick describes[58] the problem:

> "Every time radioactive waste is dumped into a stream, buried, dropped into the ocean, discharged into the air, or otherwise released from human control, it passes into the complex world of living things. It will pass from living thing to living thing, sometimes becoming concentrated, at other times being dispersed, with an efficiency and ingenuity which man has not yet come to understand. At unpredictable times and places, this radioactive waste will reappear in man's food, air and water. It will not go away for decades or centuries or even millennia."

Society has a negative image and a fear of radiation at any level; radiation is invisible and gives no indication of its presence. Society also worries about nuclear waste because radioactive wastes are extraordinarily toxic and some very long-lived. The half-life of man-made plutonium is 24,000 years. There is already an enormous quantity of such wastes mostly from the nuclear weapons development. According to Kerr,[59] more than a quarter-million cubic meters of liquid wastes are now being held at government installations awaiting final disposal. Confinement of wastes for periods of several thousand years is technically feasible and a number of methods of disposal of long-lived highly radioactive wastes seem promising.[60] Beyond 1000 years the ingestion toxicity is comparable to that of the original uranium from which the wastes were derived. Glass is regarded by many as a reasonable waste form in geologically stable salt beds, but it is difficult to find states and communities willing to host disposal facilities. In some respect nuclear waste is a millennium-long problem. Its solution requires man to plan for institutions whose stability, acceptability, and effectiveness can stretch for a millennium in the future. Society seems unwilling or uncertain as to its ability to make the necessary long-range commitments required for the proper handling of these problems. It is concerned about proliferation of nuclear materials. It makes an historical connection of nuclear power to the atomic bomb and to the first use of nuclear fission by the military as an uncontrolled explosion of unimagined destructive capacity. The stigma of the A-bomb hangs over the nuclear power reactor and so does the phantom of Hiroshima and Nagasaki and the possibility of a nuclear war. Society, is, thus, skeptical about the benefit of nuclear power at the present time when man still satisfies his energy needs largely by using other energy sources.

Turning to fusion, thermonuclear fusion reactions generate the energy of the sun and the stars. In a fusion reaction two light atomic nuclei combine in a high-energy collision to form a heavier nucleus. When the two nuclei fuse, the mass of the products of the reaction is less than the combined mass of the two original nuclei. A small amount of matter Δm is converted into a large amount of energy as is prescribed by Equation (3).

Fusion reactions became known in the 1930s before fission reactions did. They are among the most elementary and best understood nuclear processes. In principle, most of the nuclear isotopes near the lower end of the Periodic Table could combine in nuclear fusion reactions with a net release of energy. In spite of this, and in spite of the fact that efforts to harness the energy of fusion reactions for electric power began soon after their use in the H-bomb, *controlled* fusion has not as yet been achieved. The main difficulty lies in the physical conditions that must be reached to ignite and maintain energetically self-sustaining fusion reactions. For two light nuclei to fuse, they need to come very close, and to come close enough to fuse their kinetic energy must be sufficiently high to overcome their mutual electrostatic repulsion. The light nuclei of the stars overcome this mutual repulsion because of the extremely high temperatures and strong gravitational forces prevailing there.

The primary fuel for fusion is deuterium (D_2): an inexpensive stable isotope of hydrogen with one proton and one neutron making up its nucleus. Deuterium occurs in molecules of ordinary water. It replaces hydrogen naturally at an abundance of one part deuterium per 6,500 parts hydrogen. This amounts to an enormous and virtually inexhaustible supply of deuterium in the world's oceans, lakes, and rivers. In each gallon of seawater there is about half a gram of deuterium. That half gram of deuterium has a fusion energy equivalent of 300 gallons of gasoline![61] Used as a fuel, deuterium can react with itself or with other light isotopes. The most important fusion reactions involving deuterium are given in Table 2.1. Each of these reactions burns isotopes of hydrogen into helium and produces large amounts of energy from a small quantity of fuel. The deuterium-tritium reaction (shown in bold in Table 2.1) is especially significant because it has the lowest ignition temperature (about 50 million degrees). This (D-T) reaction releases about 20 million times more energy per reaction than the chemical reaction produced by burning fossil fuels, and its by-products are harmless.

Table 2.1. The most important fusion reactions involving deuterium[62]

Fusion reaction [a,b]	Energy released[c] (MeV)	Energy released[c] (kwh/g)
Deuterium (D) + Deuterium (D) → Tritium (T) + Proton (p)	3.25	22,000
Deuterium (D) + Deuterium (D) → Helium 3 (^{3}He) + Neutron (n)	4.0	27,000
Deuterium (D) + Tritium (T) → Helium 4 (^{4}He) + Neutron (n)	**17.6**	**94,000**
Deuterium (D) + Helium 3 (^{3}He) → Helium 4 (^{4}He) + Proton (p)	18.3	98,000

[a] Tritium is an unstable, heavier isotope of hydrogen (two neutrons and one proton), and helium-3 is a stable light isotope of helium (two protons and one neutron).

[b] Tritium is radioactive and decays with a half-life of 12.3 years. It is not a naturally occurring element. It can be produced by irradiating lithium with neutrons.

[c] For comparison, the energy for the chemical combustion reaction $2 H_2 + O_2 → 2 H_2O$ is 0.000006 MeV or 0.0044 kwh/g.

The fundamental problem in producing controlled fusion reactions is that matter must be heated to millions of degrees centigrade. At these temperatures, matter only exists in the form of a plasma -- a gas composed of an equal number of positively charged nuclei and free electrons. Confinement of the plasma is the central scientific issue in efforts to achieve fusion energy. The plasma fuel must be dense enough and must stay hot long enough to allow fusion reactions. The so called fusion triple product - - ion density × ion temperature × ion confinement time - - must exceed ~8 x 10^{20} m^{-3} keV s for "breakeven" and a level about five times higher for "ignition."[63] Achieved levels to date are about a factor of 3 to 5 below the breakeven level.[64] Two types of nonmaterial plasma confinements are being attempted: magnetic and inertial. In magnetic confinement the charged particles of the plasma are constrained within a defined region by intense and especially shaped magnetic fields. In a sense the magnetic field acts as a nonmaterial furnace liner that insulates the hot plasma from the material chamber walls. In inertial confinement,[64,65] a very powerful source of energy (lasers or particle beams from accelerators) is employed to deliver sufficient energy to compress and heat within a short time (billionths of a second) a small frozen pellet of fusion fuel to its ignition point. Laser-initiated fusion is hoped to be achieved by compressing and heating a tiny pellet containing deuterium and tritium by carbon dioxide or other type of laser beams. This causes the inner shell of the target to implode and compress the fuel. The compression must heat the fuel to at least 50 million degrees and compress it to densities a thousand times greater than normal within nanoseconds.[64,65] Much of the technical complexity of fusion arises from unwanted cooling of the fuel and limitations in achieving sufficient increases in plasma density stemming from instabilities created by the external forces used to compress the plasma. The workability of the magnetic confinement and the inertial confinement using lasers is broadly accepted, but the generation of power from fusion reactors is still at an unpredictable time in the future.

In closing this brief discussion on energy, we restate the obvious: man needs every energy source he can develop, and every energy source he develops has its benefits and its problems. In particular, nuclear power from fission is still faced with the issues of reactor safety, waste disposal, transport of radioactive materials, proliferation of radioactive materials, and public acceptability. Controlled nuclear fusion may well be man's future energy source. Fusion energy is thought to be safe, clean, cheap, and inexhaustible. Its further development would require scientific research and insight, new technology and new materials, and people with expertise and purpose. Both nuclear power from fission and future nuclear power from fusion, require human institutions with longevity and stability to exert "eternal" vigilance for the proper and safe operation of man's nuclear energy system. In the proper development and control of the new sources of energy lies man's survival and freedom, in their absence or abuse the absence of both.

2.5. THE POWER OF THE GREEK MYTH

Let us close this chapter by referring to four interrelated Greek myths which are relevant to the present discussion on the science and technology of modern man. The myths of Prometheus and Pandora, and those of Daedalus and Ariadne.[66-68]

2.5.1. Prometheus and Pandora

Prometheus (meaning "the foreseeing," "before learning") was a titan[69] who stole fire from the gods, carried it to earth, and taught mankind its use. The theft brought down upon Prometheus the wrath of Zeus and a severe punishment: Prometheus was chained to a rock where each day a vulture came to eat his liver away, which was made up again each night to be eaten away yet again each and every successive day. An unending torture and a perpetual agony. A severe punishment for a severe crime.

Zeus plotted vengeance against Prometheus and punishment of mankind for Prometheus' gift. He caused to be created a woman, whom the gods endowed with beauty, "treacherous nature," and a heart filled with lies and "wheeling words of falsehood" to beguile Prometheus to his ruin and to be a sorrow to men.[67] This woman was called Pandora (meaning "all gifts," "the all-giver"). Pandora had a box containing all manner of evils and diseases which she opened, and they all flew out, except *hope* ($\epsilon\lambda\pi\iota\varsigma$). Only *hope* remained in Pandora's box, only it remained under man's control.

Modern science is likened to Prometheus. Like Prometheus, the modern scientist handed modern man a new fire, a cosmic fire, *the nuclear power.* This power has chained humanity to the rock of perpetual agony: life will for ever be under the shadow of catastrophe. For it is possible for man to burn himself to death. It is also possible for modern man to change fundamentally himself through the enormous powers -- the new types of primordial fires -- modern science has put in his hands especially in the areas of genetic engineering.

Modern science is also likened to Pandora. It opened up the scientific and technological box like Pandora did hers. The box of scientific knowledge is and will remain open for ever, like Pandora's. And mankind is not ready to handle such enormous powers.

Powerful myths and chilling analogies. But, man, unlike any other animal, learned to use fire. As it was with the mythological Pandora, so it is with the modern one: man still has hope.

2.5.2. Daedalus and Ariadne

Daedalus was a great craftsman, an expert, possibly the first Greek inventor. From professional jealousy he killed his colleague Talos and was exiled from Athens. He went to Crete to work for king Minos. There, Daedalus' ingenuity solved a number of problems of questionable moral value.[67,70] He constructed a machine (a wooden cow) that enabled Mino's wife to copulate with a bull. From their union was born a monster, half-bull, half-man, the Minotaur. To help Minos hide away the monster, Daedalus devised and built the Labyrinth with its innumerable maze-like passages, and placed the Minotaur in its depths. He, also, invented a device for solving the puzzle of the Labyrinth. Daedalus was useful to Minos and for this reason he refused to allow him to leave. Daedalus solved this problem by inventing flying that carried him to Sicily. Minos valued Daedalus' services and for this reason he went after him. When he found him he wanted to take him back. Daedalus, who did not wish to return to Crete, constructed a bath in such a way that when Minos pulled the shower handle he was scalded to death. On his way out of Crete, Daedalus took with him his son Icarus for whom he also built a flying machine using feathers, linen, and wax. He instructed Icarus to fly and cautioned him not to fly too high for the sun may melt the wax and he then would fall to his death. Icarus initially followed Daedalus' instructions but, in his exhilaration, became overconfident and daring and flew higher and higher, closer and closer to the sun. Too close to the sun, the wax melted and Icarus fell to his death in the Aegean sea.

Ariadne was the daughter of king Minos and queen Pasiphaë. She met the Athenian Theseus, who volunteered to come to Crete as one of a yearly tribute of seven youths and seven maidens to Minos by Athens. The tribute Minos shut up in the Labyrinth, to lose their way and die of hunger, or be killed by the Minotaur. Ariadne fell in love with Theseus and gave him a ball of thread to unwind when he entered the Labyrinth. Theseus met and killed the Minotaur, and by following the clue that Ariadne gave him, i.e., by winding the thread up again, found his way back to the entrance of the Labyrinth and escaped taking with him his fellow victims and Ariadne.

Fascinating and powerful myths that triggered innumerable analogies to subsequent situations in human history. We have the image of Daedalus as a man passionately devoted to his work and inventions, a man challenged by the thrill of finding solutions to problems with little concern about the uses and implications of his inventions. An expert in mad pursue of a device for whomever happens to be his "customer." A jealous professional who migrates wherever he perceives the challenge to be and wherever he can best do what he wants to do. A skillful man valued for his use by people in power. A self-confident and disloyal employee, one who attaches no moral value to the uses of his inventions, one who instills confidence and entrusts his powerful inventions to people who are overconfident and unskilled to fully realize the consequences of the inventions, and whose overconfidence inevitably leads them to disaster. Bertrand Russell expressed[71] such pessimistic view: "I fear," he said, "that the same fate (like Icarus's) may overtake the populations whom modern men of science have taught to fly." As for Ariadne, it was her, not Daedalus, who helped the victims of Minos' brutality. Daedalus just devised the scheme to exit the Labyrinth. It was Ariadne -- not Daedalus -- who made the moral decision and sided with the victims. It is with Ariadne's string that man can find his way out of modern Daedalus' Labyrinth.

The fire and the Labyrinth are the works of the scientist and the technologist, but the hope and the way out of the Labyrinth can only partly be the work of science and technology. They must come from the will and the values of man, the timeless. And man must succeed. There is no room for tragedy. Change has changed that as well. Tragedy has a hero who is intrinsically good, but who destroys himself without knowing it, just like Oedipus. Modern man should not be ignorant of the consequences of his failing.

2.6. REFERENCES AND NOTES

1. T. C. McLuhan, *The Way of the Earth*, Simon & Schuster, New York, 1994, p. 270.
2. D. Bell, quoted by G. T. Seaborg, in *The Future of Science, 1975 Nobel Conference*, T. C. L. Robinson (Ed.), Wiley-Interscience, New York, 1977, p. 10.
3. The membership of the United Nations grew from 51 member states when it was formed in 1945 to 130 by 1971, and to 185 by the mid 1990s.
4. M. C. Goodall, *Science, Logic, and Political Action*, Schenkman Publishing Co., Inc., Cambridge, MA, 1970, p. 54.
5. It was Albert Einstein's letter to President Franklin D. Roosevelt which led to the Manhattan Project and the atomic bomb. Here is part of what the letter said (J. Ziman, *The Force of Knowledge*, Cambridge University Press, Cambridge, 1976, p. 128):

   "Some recent work by E. Fermi and L. Szilard, which has been communicated to me in manuscript, leads me to expect that the element uranium may be turned into a new and important source of energy in the immediate future. Certain aspects of the situation which has arisen seem to call for watchfulness and, if necessary, quick action on the part of the Administration. I believe therefore that it is my duty to bring to your attention the following facts and recommendations:
   - In the course of the last four months it has been made possible -- through the work of Joliot in France as well as Fermi and Szilard in America -- that it may become possible to set up a nuclear chain reaction in a large mass of uranium, by which vast amounts of power and large quantities of new radium-like elements would be generated. Now it appears almost certain that this could be achieved in the

immediate future.
- This new phenomenon would also lead to the construction of bombs, and it is conceivable -- though much less certain -- that extremely powerful bombs of a new type may thus be constructed. A single bomb of this type, carried by boat and exploded in a port, might very well destroy the whole port together with some of the surrounding territory...."

By most accounts, the famous letter was written by Leo Szilard and Eugene Wigner who then convinced Einstein to sign it. However, according to David Sundberg (The Oak Ridger, March 11, 1986, Oak Ridge, TN), Wigner's account is different. Here is how Sundberg described Wigner's account in an interview with him:

"Leo Szilard who, like Wigner, was born in Hungary, and Wigner made unsuccessful attempts, following the discovery of fission in Germany in 1939, to interest the U.S. government to undertake an atomic 'chain reaction' program. Their initial appeals fell on deaf ears. Then Szilard suggested that the two of them approach Einstein, perhaps the only scientist in America whose reputation was sufficient to convince the President to pursue nuclear weapons development. Wigner and Szilard went to Einstein's summer home on Long Island and spoke to him in German because of the famed mathematician's poor grasp of English. Einstein may have heard about the announcement of the fissioning of the uranium nucleus shortly before the conversation, but clearly had not heard of the principles involved. Wigner said, that, within 15 minutes, Einstein 'understood it, saw the danger, and dictated a letter which I took down in German, took it back to Princeton, translated it, and had it typed and he signed it."

6. H. Jonas, *The Imperative of Responsibility*, The University of Chicago Press, Chicago, Il, 1984, p. 5.
7. The first atomic bomb was tested by the USA in a desert of Southern New Mexico on July 16, 1945. The Soviet Union exploded its first atomic bomb on August 29, 1949. Britain, France, and China followed, respectively, on October 3, 1952, February 13, 1960, and October 16, 1964. The first hydrogen bomb (or fusion bomb, or thermonuclear bomb) was exploded by the USA on October 31,1952 at Eniwetok in the Marshal Islands. The Soviet Union, Britain, France, and China, exploded their fusion bombs respectively on November 22, 1955, November 8,1957(?), August 24, 1968, and June 17, 1967. India conducted its first five nuclear underground tests on May 11 and 12, 1998, and Pakistan its first five nuclear underground tests all on May 27, 1998.
8. D. Albright and K. O'Neill, The Bulletin of the Atomic Scientists, January/February 1995, p. 20.
9. T. Reed and A. Kramish, Physics Today, November 1996, p. 31.
10. H. Friedman, L. B. Lockhart, and I. H. Blifford, Physics Today, November 1996, p. 38.
11. D. Hirsch and W. G. Mathews, The Bulletin of the Atomic Scientists, January/February 1990, p. 2.
12. D. Albright, The Bulletin of the Atomic Scientists, June 1993, p. 14.
13. E. Segré, *From X- Rays to Quarks,* W. H. Freeman and Co., New York, 1980, p. 217.

14. R. S. Norris, The Bulletin of the Atomic Scientists, November/December 1994, p. 58.

15. The atomic bomb dropped on Hiroshima had an explosive force of about 12,000 tons of TNT or 0.012 MT (F. Barnaby, in *Trends in Physics*, M. M. Woolfson (Ed.), Adam Hilger Ltd., Bristol, 1978, p. 51). It was made of uranium 235. The second bomb which was dropped on Nagasaki was made of plutonium.

16. R. P. Turco, O. B. Toon, T. P. Ackerman, J. B. Pollack, and C. Sagan, Science **222**, 23 December 1983, p. 1283; Science **247**, 12 January 1990, p. 166; L. Sartori, Physics Today, March 1983, p. 32.

17. D. A. Hamburg, in *Science, Technology, and Society*, R. Chalk (Ed.), American Association for the Advancement of Science, 1988, p. 114.

18. The world came close to experiencing a nuclear exchange during the Cuban missile crisis in 1962. M. Moore (The Bulletin of the Atomic Scientists, November/December 1995, p. 16) states that "In the Korean War, Eisenhower hinted broadly that he might use nuclear weapons to bring the war to an end." Similarly, E. R. May, (*"Lessons" of the Past*, Oxford University Press, New York, 1973, p. 107) refers to a meeting on the planning of the Vietnam War where the representative of the Joint Chiefs is quoted as having said "Possibly even the use of nuclear weapons at some point is of course why we spend billions to have them." The United States would have seriously considered using nuclear weapons against Iraq, if Iraq used biological weapons of mass destruction, according to former USA Secretary of Defense R. B. Cheney (CNN with B. Shaw, February 27, 1996). Nuclear weapons can be made to be purpose specific (e.g., the neutron bomb) and thus make "easier" the decision to use them. The possibility also exists that such terror weapons may go off accidentally, or that they may find themselves in the hands of terrorists, or that they may be under the loose control of states not equipped to secure their custody.

19. Science **269**, 28 July 1995, p. 483.

20. E. Rabinowitch, Bulletin of Atomic Scientists XII, January 1956, p. 2.

21. F. Dyson, *Weapons and Hope*, Harper & Row, Publishers, New York, 1984.

22. H. F. York, Physics Today, April 1988, p. 40.

23. D. J. Bradley, C. W. Frank, and Y. Mikerin, Physics Today, April 1996, p. 40.

24. Although nuclear weapons are not easy to be used by terrorists, safeguarding fissile material in the post-cold-war era is worrisome (see, for example, F. von Hippel, Physics Today, June 1995, p. 26).

25. I. Goodwin, Physics Today, July 1997, p. 47.

26. R. Dubois, *A God Within*, Charles Schribner's Sons, New York, 1972, Chapter 7.

27. V. F. Weisskopf, Physics Today, July 1978, p. 32.

28. Much has been written about the biological effects of nuclear testing and the concerns that such tests have not been carried out "with disciplined scientific procedures." Between 1945 and December 1992, there have been a total of 1950 known nuclear tests worldwide (R. S. Norris, The Bulletin of the Atomic Scientists, April 1993, p. 48). About 85 percent of these were exploded by the USA and the Former Soviet Union and about 27 percent of all tests, i.e., 525 nuclear explosions, were atmospheric. Also, significant amounts of radiation have been released in the environment as a result of nuclear weapon development, mostly by the USA and the Former Soviet Union. The cause of most environmental contamination has been the reprocessing of nuclear fuel from reactors which were used to produce weapons materials. According to Bradley et al. (Ref. 23), the total amount of radioactivity released by the USA and the Soviet

Union is "roughly 1.7 billion curies in current radioactivity" (one curie is 3.7×10^{10} nuclear disintegrations per second). This radioactivity is concentrated in small areas.

29. Industrial pollution, especially atmospheric halocarbons, threatens to destroy (decrease) the amount of ozone (O_3) of the stratosphere. Should this happen, this protective layer -- protective in that it absorbs ultraviolet radiation which can be harmful to life on earth -- will no longer effectively filter the ultraviolet radiation.

30. The earth's surface emits infrared radiation a portion of which is absorbed and reradiated back by atmospheric gases such as CO_2, H_2O, CH_4 and N_2O which have strong infrared absorption, especially in the wavelength range from about 7 μm to 13 μm. Foremost among these naturally occurring effective infrared-radiation-trapping or greenhouse gases is carbon dioxide (CO_2) which principally regulates the earth's temperature. To the naturally occurring greenhouse gases man adds large quantities of greenhouse gases (e.g., CO_2) through the combustion of fossil fuels and other man-made gases (e.g., perfluorocarbons). These man-made greenhouse gases cause an enhancement in the greenhouse effect above and beyond the level due to the naturally-occurring greenhouse gases. Man-produced greenhouse gases, especially the additional amount of CO_2, shift the balance between the incoming and outgoing radiation at the top of the troposphere toward the former causing global warming. It is feared that the enhanced greenhouse effect due to the increased amount of CO_2 and other man-made greenhouse gases in the environment can cause the earth's temperature to rise.

31. F. Joos, Europhysics News **27**, 213 (1996).

32. M. Perutz, *Is science Necessary?*, Oxford University Press, Oxford, 1991, p. 24.

33. A. M. Weinberg, Minerva **10**, April 1972, p. 209; Science **177**, 21 July 1972, p. 211.

34. EPRI Journal, October/November, 1987 and March 1992; Physics Today, January 1995, p. 13; M. S. Linet and others, The New England Journal of Medicine **337**, July 1997, p. 1; *Assessment of Health Effects from Exposure to Power-Line Frequency Electric and Magnetic Fields*, NIH Publication No. 98-3981, August 1998, National Institute of Environmental Health Sciences, U.S. National Institutes of Health, U.S. Department of Health and Human Services, P.O. Box 12233, Research Triangle, NC 27709.

35. R. Dubois, *A God Within*, Charles Schribner's Sons, New York, 1972, pp. 148,149.

36. R. Dubois, *Reason Awake*, Columbia University Press, New York, 1970, pp. 167, 168.

37. This is clearly the teaching of the parables of the talents (Luke 19, verses 12-26, Matthew 25, verses 14-30), and the workers of the vineyard (Luke 20, verses 9-16). It is also the pastoral spirit of the parable of the Good Shepherd (Luke 15, verses 4-7).

38. Science **194**, 12 November 1976, p. 704.

39. F. Cramer, in *Scientists in Search of Their Conscience*, A. R. Michaelis and H. Harvey (Eds.), Springer-Verlag, Berlin, New York, 1973, pp. 19-30.

40. H. Geissbühler, P. Brenneisen, and H.-P. Fischer, Science **217**, 6 August 1982, p. 505.

41. T. N. Barr, Science **214**, 4 December 1981, p. 1087.

42. N. C. Brady, Science **218**, 26 November 1982, p. 26.

43. G. J. V. Nossal, in *The Greatest Adventure*, E. H. Kone and H. J. Jordan (Eds.), The Rockefeller University Press, New York, 1974, p. 153.

44. Of the approximately 250,000 plants known to man, fewer than 100 are used on any large scale for food, and only about a dozen or so provide directly or indirectly 90% of the world's food supply [G. T. Seaborg, in *The Future of Science,1975 Nobel Conference*, T. C. L. Robinson (Ed.), Wiley-Interscience, New York, 1977, p. 15; Science **181**, 6 July 1973, p. 13].

45. V. E. McKelvey, in *The Greatest Adventure*, E. H. Kone and H. J. Jordan (Eds.), The Rockefeller University Press, New York, 1974, p. 209, 210.

46. C. Starr, M. F. Searl, and S. Alpert, Science **256**, 15 May 1992, p. 981.

47. E. Cook, *Man, Energy, Society*, W. H. Freeman and Company, San Francisco, 1976, p. 1.

48. J. R. Maddox, *Beyond the Energy Crisis*, McGraw-Hill Book Company, New York, 1975, p. 34.

49. L. Motz and J. H. Weaver, *The Story of Physics*, Plenum Press, New York, 1989, p. 164.

50. B. Commoner, *The Poverty of Power*, Alfred A. Knopf, New York, 1976, p. 28.

51. J. P. McBride, R. E. Moore, J. P. Witherspoon, and R. E. Blanco, Science **202**, 8 December 1978, p. 1045.

52. M. Calvin, Science **219**, 7 January 1983, p. 24; P. H. Abelson, Science **191**, 26 March 1976.

53. J. L. Stone, Physics Today, September 1993, p. 22.

54. D. Bodansky, Science **207**, 15 February 1980, p. 721.

55. C. P. Zaleski, Science **208**, 11 April 1980, p. 137.

56. A. Wyatt, *The Nuclear Challenge, Understanding the Debate*, The Book Press Ltd., Toronto, Canada, 1978.

57. A. M. Weinberg, The Bulletin of the Atomic Scientists, March 1980, p. 31.

58. S. Novick, *The Careless Atom*, Houghton Mifflin Company, Boston, 1968, p. 103.

59. R. A. Kerr, Science **204**, 20 April 1979, p. 289.

60. J. M. Harrison, Science **226**, 5 October 1984, p. 11; Physics Today, June 1997.

61. Ref. 47, p.71.

62. R. F. Post and F. L. Ribe, Science **186**, 1 November 1974, p. 397.

63. R. R. Parker, Journal of Fusion Energy **10**, 83 (1991).

64. Journal of Fusion Energy **10**, 83 (1991).

65. J. Nuckolls, J. Emmett, and L. Wood, Physics Today, August 1973, p. 46; J. H. Nuckolls, Physics Today, September 1982, p. 24.

66. M. Grant, *Myths of the Greeks and Romans*, Mentor/Penguin Books USA Inc.,New York, 1962.

67. H. J. Rose, *A Handbook of Greek Mythology*, E. P. Dutton & Co., Inc., New York, 1959.

68. R. Graves, *The Greek Myths*, Penguin Books, London, 1955.

69. Titans were giant deities of Greek mythology who were overthrown and succeeded by the Olympian gods of ancient Greece.

70. S. Dedijer, Science **133**, 30 June 1961, p. 2047.

71. B. Russell, *The Future of Science*, Philosophical Library, New York, 1959, p. 2.

CHAPTER 3

The Way Science Works and Evolves

There are many complementary ways to the truth and science is one of them. Learning began long before science, but such learning, as far as the natural world is concerned, was primitive and qualitative. Man has lived without science for most of his life.

Science is founded on the postulate that nature is rational. It is based on man's deep-rooted need to know and on his urge to improve his condition through such knowledge. Man by his nature wants to know maintained Aristotle. The word *science* itself evolved from the Latin *scio* (meaning "I know"). It was introduced at the beginning of the 17th century and was synonymous with knowledge. By the mid-19th century science came to represent a particular kind of knowledge derived deductively or through observation and experiment. According to Dubois,[1] the word *scientist* appeared in print in 1841 and William Whewell is credited[2] as having introduced it. Prior to that time the terms "Man of Science" or "Natural

Philosopher" were in use. In Webster's dictionary science is defined as knowledge. And knowledge rests upon other knowledge, it is irreversible and a prerequisite of truth. Russell[3] divides science into three groups: physical, biological, and anthropological with the last group embracing all studies concerned with man (human physiology and psychology, anthropology, history, sociology, economics, and medicine). Most of our discussion will be on physical science.

Science is deductive and inductive, it concentrates on similarities and on organized and systematic observation and experimentation. It is by inductive reasoning that science normally discovers, and it is by deductive reasoning that science normally provides intellectual proof. Scientific knowledge is dynamic and emerging. Although it is rational, it evolves. It is tentative and flexible. It is always in a state of a continuous readjustment. It describes nature the way nature appears to be, not necessarily the way nature is. Due to its transitory character, scientific knowledge does not furnish ultimate explanations. It has been and will remain incomplete. The laws of science are mostly inductive generalizations representing enormous compression of information, succinct and precise quantitative relationships found by repeated experiments reflecting persistent regularities in the behavior of the physical world, but they are not laws for everything. Physical laws are human constructs and as such are limited by human understanding and action. They are not rigid. Time and again one law leads to and is superseded by another more general and more precise for the truth that science searches for is not timeless. Science discovers and describes the existing, not the essence of reality. It answers to the question "how?", not to the question "why?". It tells us, for example, that the uniformity and the validity of the laws of nature are universal but it does not tell us why this is so. Science, however, as a "means of establishing new kinds of contacts with the world, in new domains, on new levels,"[4] as a system of validating knowledge, as a perpetual search for the regularities, the relationships, the commonalities, the order, the causes and the laws which govern natural processes and phenomena, as the unraveller of the "hidden likeness"[5] in nature -- and thus as the unifier of man's knowledge of the physical world -- possesses the power of not only to understand, but also to predict and to control natural events and processes.

Thus, traditionally, science has had two main functions. First, acquisition of new knowledge about the physical world including the organization and the codification of existing knowledge. Second, application of knowledge for practical purposes. Both of these functions are conducted by and for people. Therefore, "natural science always presupposes man."[6] Science is first and foremost a *human activity* and a *social process*. Men and women are its practitioners. The educational and the research institutions of man are its main arenas. Social, economic, political, intellectual and even moral institutions fall within its sphere of influence. Written and unwritten norms govern (should govern) the mechanisms of its operation: the production, diffusion, and application of scientific knowledge.

3.1. THE SCIENCE OF GREEK ANTIQUITY

Modern science, it is said,[7] is distinctly European (Western). Science is, however, old. Its roots lie deep in several ancient cultures such as those of China, Egypt, Asia Minor, and foremost Greece. The origin of science can be traced to the observation of the heavens, the stars, the regular motion of the planets, and the occurrences of natural phenomena.[8] At the origin of science lie magic, astrology, and religion. The majestic beauty and the mysterious powers of the heavens must have overwhelmed our ancestors and must have instilled in them awe and fear, reverence and curiosity, the idea of a fateful control of their destiny by the imposing almighty heaven.

The astrologers of Egypt and Babylon kept empirical records, had units and rules of measurement, developed a calendar of the year, and recognized the periodicity of astronomical events.[9,10] However, it is the Greek philosophers of antiquity who first submitted such knowledge to rational examination and attempted to trace causal relations among its parts, that is, to create the first science. It is they who principally wondered about "the substratum of nature"[11] and believed that it is possible to reach a rational explanation of the cosmos. The story of physics begins with Greek science and Greek science began with astronomy and with the Greek natural philosophers of Ionia around the 7th century B. C. It stretched for about six centuries, through the golden age of classical Greece (5th and 4rth century B. C.). Altogether, Greece's science was linked to philosophy and to pure thought and was supremely deductive in character. But it was also observational and empirical. Detached from application and experiment[12] -- and deprived of correlation and mathematical relationship -- Greek science had no predictive capacity. To Plato (427-347 B. C.), thinking was man's highest activity. Nonetheless, the embryology of science -- the search for simple principles underlying the observed phenomena -- and many scientific dreams began here. Examples of ideas of the Greek antiquity that impacted science's subsequent evolution are sketched below.

Astronomy the Forerunner of Classical Physics. Greek astronomers accumulated systematically a vast amount of astronomical data. This database allowed them to formulate the first ideas about the earth and the solar system, which triggered -- centuries later -- the science of astronomy, kinematics, dynamics, and eventually classical mechanics. They developed the geocentric (earth-centered) system of the universe (Pythagoras, 560-480 B. C.) and yet amongst them was the dissenting view of Aristarchus of Samos (310-230 B. C.) who believed that the solar system is not geocentric but heliocentric (sun-centered); it took two thousand years to prove the correctness of Aristarchus's hypothesis! Pythagoras (560-480 B. C.), as well as Aristotle (384-322 B. C.) and Plato, envisioned the heavens rotating around a *fixed* earth and yet amongst them was the dissenting Herakleides (4rth century, B. C.) who's idea was one of a *revolving* earth. And clearly the Greeks showed the usefulness of mathematics by calculating such things as the diameter of the earth. They might have even thought of the idea that astronomical objects obey unseen forces. However, neither the idea of using mathematics to provide a common language for the sciences, nor the importance of a force at a distance were seriously considered by them. Consequently, the Greeks were unable to move beyond observation to the formulation of the physical law. They did not ask the physics question: "What makes the planets move the way they do?" Had they have done so, they would have been confronted with the question of the forces exerted on the macroscopic bodies that would have explained their movements and orbits. And yet Aristotle envisioned the universe as made of matter and forces -- as we do today. They can take comfort in that their observational and "theoretical" work led to this question 2000 years later. Only the path -- not the solution -- can be traced to their science. This path was enriched by the work of Hipparchus (ca 190-120 B. C.), the greatest astronomical observer of ancient Greece, and was illuminated by the work of Ptolemy (100-170 A. D.), the great synthesizer of the work of Greek astronomers. Ptolemy's book, *Almagest*, was the European astronomer's textbook for 1500 years.

Geometry, Morphology, and Mathematics. The Greeks were exceptional geometers, and geometry is important to science. It is, for example, in a geometrical context that the laws of motion of bodies can be expressed. Geometry best expresses morphology and connects science to art. If God is a geometer -- as Plato thought -- and if knowledge of geometry was the necessary prerequisite to enter Plato's Academy, then it is Euclid of Alexandria (3rd century B. C.) who proved Plato's argument. His book, *Elements*, summarized the mathematical knowledge of ancient Greece, 13 volumes of definitions, axioms, and theorems! It served for centuries as the correct geometrical framework on which the laws of nature were formulated. Newtonian mechanics and Maxwell's electromagnetism incorporated Euclidean geometry into their theoretical formalisms. Greek mathematics was general in its premise and strictly logical in its reasoning. It finds its expression in Pythagoras and his school. Einstein continued the Pythagorean tradition 25 centuries later when he developed his ideas on relativity.

The Concepts of Continuity, Atomicity, Duality, and Complementarity. Heraclitus (ca 500 B. C.) is perhaps the first classical physicist. He maintained that the law of the cosmos is change. Everything is in a continuous state of flux, everything is undergoing a perpetual change, everything is an "eternal becoming," ceaseless energy. In diametric opposition to Heraclitus's view, Zeno of Elea (5th century B. C.) maintained that nothing moves, and Parmenides of Elea (5th century B. C.) considered change an illusion. Parmenides perceived the world as an indestructible substance, an "unchangeable being." The *being* is (εστίν) and hence it is conserved (interestingly, this may well be the first explicit conservation principle![13]). To Parmenides, the basic ontological structure is the *unity of being*, the unity of the world, the whole, the unchanging reality. These extreme and diametrically opposing views were reconciled by the philosophers/scientists of the 5th century B. C. in the most extraordinarily ingenious way: that of *complementarity*. The being, they said, is manifest in certain *invariable* substances the mixture and separation of which gives rise to the *changes* in the world. Hence, the concept of the *atom*, the smallest indivisible and unchangeable unit of matter which found its clearest expression in the philosophy of Leucippus (445 B. C.), Democritus (460-357 B. C.), and, later, Lucretius (98-55 B. C). The atoms, the Greek atomists prescribed, were passive, dead particles, moving in a void; they differed in many ways (e.g., size, mass, color), and combined with each other to form the observable matter. Matter was made of these "building blocks," the atoms. The cause of the atomic motion was not explained, but was often associated with external sources which were assumed to be spiritual in origin and fundamentally different from matter. Thus, a clear line was drawn between matter and spirit. Out of these philosophical discussions and conceptions was born the theory of atomic constitution of matter and the concept of dualism. The former -- the possibility that matter is built up of unchangeable individual units, the atoms -- lay dormant until the dawn of physics and chemistry 2300 years later at the end of the 18th century when Lavoisier and Dalton took over from the Greeks. The latter -- the idea of complementarity between the "eternal becoming" and the "unchangeable being" -- emerged into the dualism between mind and matter -- body and soul (spirit) -- and became an essential element of Western Culture. And there was still Anaxagoras (ca 500-428 B. C.) who argued that things undergo unlimited division.

The Greeks knew about electricity and magnetism (Thales of Miletus first observed static electricity), but did not relate their knowledge on electricity and magnetism to the atom of Democritus. It was too early for that.

The Method of Scientific Inference and Reflective Reasoning. To Plato, God was a geometer. God's havens, therefore, should exhibit perfect geometrical shapes, i.e., the heavenly bodies had to move in circles, a belief that carried through to Kepler 2000 years later. This is just an example of *deductive* reasoning: God created the heavens, God is perfect, hence the orbits of the heavenly bodies must be perfect, hence circular. The Greeks, foremost among them Plato, obtained their models in this manner -- deductively -- by inference from some fundamental axiom (i.e., a self-evident, accepted proposition), the "paradigm" of modern science. Deductive reasoning has been a basic element of human thinking since. And although, as we shall see later, the method of modern science is basically inductive rather than deductive reasoning, great scientists triumphed through the latter. Einstein, for example, 2300 years after Plato, postulated the constancy of the speed of light and simply looked for the consequences of such a postulate: he was led to the quantum theory of light and special relativity.

The Concept of the University, the Mission-Oriented Institution, the Scientific Advisor. The Greeks established institutions of learning that might be identified as the first precursors of today's university: Plato his Academy and Aristotle his Lyceum. Plato's Academy was devoted to mathematics and political philosophy and can perhaps be looked at as the first mission-oriented institution (its mission was to produce "the benevolent king" to run the State). Aristotle's Lyceum emphasized biology and the natural sciences, the primacy of experience (εμπυρεία), and the value of observation. Plato was the ultimate conceptualizer of the world of perfect forms, a refiner of ideas. Aristotle -- unique among his contemporaries -- was the patient observer of the infinite detail of the physical world. Aristotle's science was broad, ranging from physics to astronomy, from meteorology and geology to biology and psychology. His physics was, however, mostly metaphysics. For him the cosmos was arranged as it was pre-designed and pre-determined by the "Unmoved First Mover." Aristotle's voluminous writings, *On Physics*, influenced human thought for over 2000 years. It can be argued that he has been the most influential scientist ever lived. Aristotle was the tutor of, and later the scientific adviser to, Alexander the Great. The first scientific adviser! A role model indeed!

The First Scientific Syntheses and the First Scientific Synthesizers. The science of Greek antiquity had been synthesized, preserved, and transmitted to future generations by the first three great synthesizers of science: Aristotle, Euclid, and Ptolemy. Their respective treatises -- *On Physics, Elements*, and *Almagest* -- summarized, respectively, Greece's contributions to general science, mathematics (foremost geometry), and astronomy. They are perhaps the first classic examples of *scientific synthesis:* careful gathering and validation of observations and facts brought into a synthetic whole, an assessed synthesis, and a preservation of a sound scientific database, ideas, and understanding for the next level of science in its perpetual evolution.

Then and Now. How things have changed since can be seen from the kind of questions the scientists/philosophers of that time were preoccupied with compared to those science is preoccupied with today: general questions with limited, incomplete, and perhaps not possible answers then; restricted questions with complete or nearly complete answers today. Ontological kind of questions then, purely materialistic, factual questions now. Let us then look at Table 3.1 where we list the questions on the agenda of an old symposium held at Corinth, Greece, long ago (6th century B. C.) and described years later by Plutarch,[14] and the answers given to them. The symposium -- possibly the first "scientific" symposium -- was attended by the seven wise men of ancient Greece to consider the questions posed by the then learned king of Egypt Amasis. These same questions king Amasis sent earlier to the king of Ethiopia seeking his answers. The Greek scientist/philosopher Thales of Miletus (640?-546 B. C.), the man who introduced geometry into Greece, answered on behalf of the symposium's participants the same questions the king of Ethiopia responded to earlier. The questions of the king of Egypt, the answers of the king of Ethiopia, and the answers of Thales of Greece speak clearly of the outlook maintained by each of the three cultures they represented, but they also show how different the nature of scientific inquiry was then compared to today. By

Table 3.1. Questions and answers at an old symposium[14]

Questions (King of Egypt)	Answers (King of Ethiopia)	Answers (Thales of Greece)
Τί πρεσβύτατον; What is the oldest thing?	Χρόνος Time	Θεός God
Τί μέγιστον; What is greatest?	Κόσμος The Universe	Χῶρος Space
Τί κάλλιστον; What is most beautiful?	Φῶς Light	Κόσμος The universe
Τί σοφώτατον; What is wisest?	Αλήθεια Truth	Χρόνος Time
Τί κοινότατον; What is most common?	Θάνατος Death	Ελπίς Hope
Τί ωφελιμώτατον; What is most helpful?	Θεός God	Αρετή Virtue
Τί βλαβερώτατον; What is most harmful?	Δαίμων Demon (An evil spirit)	Κακία Vice
Τί ισχυρότατον; What is strongest?	Τύχη Luck (Fortune)	Ανάγκη Necessity
Τί ρᾶστον What is easiest?	Ηδύ Pleasure	Φυσικόν The natural

way of example, we list in Table 3.2 the titles of some of the invited papers at the 1996 meeting of the American Physical Society (APS) at St. Louis, Missouri, USA. While the Greeks, the Egyptians, and the Ethiopians knew nothing of the topics that occupied science at the APS 1996 meeting -- and hence they were not supposed to have answers for them -- it is good for the modern scientist to ponder over the questions on the agenda of that old symposium. For even today, the deeper we penetrate into the structure of matter, the expanses of the universe, the complexities of living organisms, life, and man's behavior, the closer we come to the essential problems of philosophy. What is the origin of matter? What is the significance of the particles and sub-particles of which matter is composed? What is the structure and the history of the universe? How does a growing organism develop its complex structure? What is the origin of life? How do we know? Who are we? Questions, enormously significant questions. Questions which may have little to do with the immediate material needs of society, but which are central to us as humans!

Table 3.2. Topics on the agenda of just one recent meeting in just one branch of physics (solid state/material science) of the American Physical Society (March 1996 Meeting, St. Louis, Missouri, USA)[15]

Topics
Transport phenomena in quantum dots
Theory of quantum impurities
Quantum measurements and standards
Microholes and loops in superconductors
Peak effect and vortex lattice melting in high-T_c superconductors
Electronic and magnetic correlations in Mott-Hubbard insulators
Single-electron devices above 4 K
Rigid-rod and liquid-crystalline polymers
Optically-assisted nuclear magnetic resonances
Ultra fast dynamics in photoluminescence in conjugated polymers
Molecular manipulation using scanned probe techniques
Chemical sensitivity in scanning probe microscopy
X-ray correlation spectroscopy
Time-resolved electron chemistry

3.2. FROM GREEK ASTRONOMY TO THE CLASSICAL PHYSICS OF THE MACROCOSMOS

The evolutionary journey from the embryologic science of ancient Greece to the development of the first branch of physics, classical mechanics, took about 2000 years. The Romans who followed the Greeks were interested in technology rather than in science, and Byzantium and The Middle Ages saw little in scientific advances. It wasn't until the 16th century in Renaissance Europe,[16] that science took off practically from where it had stopped: Greek astronomy. Sixteenth-century Europe discovered the *experimental method*, a powerful new way to the "truth" that challenged authority, influenced the emerging movements of European liberalism, and kindled the dawn of the ascent of science. The new method offered a way to systematically test hypotheses and theories and to validate observations and deductions by direct interrogation of nature.

The spirit of 16th century Europe is personified in Nicolaus Copernicus (1473-1543). He challenged Ptolemy's geocentric planetary system and proposed instead a heliocentric one. He dethroned the earth as the center of the universe, just as Aristarchus of Samos first conceived. While Copernicus continued to accept Ptolemy's belief that the planets move in circular orbits, his rejection of the established doctrine is the first major scientific advancement since Greek science and is regarded as the embryonic moment when the scientific chain reaction began. It was followed by the abundant, systematic, and accurate (as accurate as the detection instrument of that time - - the naked human eye - - could make) astronomical observations of Tycho Brahe (1546-1601), which led Johannes Kepler (1571-1630) to his three empirical laws of planetary motion and the demonstration that the planetary orbits are not circular - - as the Greeks and Copernicus accepted them to be - - but elliptical, and thus "not perfect." Kepler was able to discover his laws only because Brahe made accurate observations. In fact, this may be the first triumph of the demonstrated value of the experimental fact (in this case observational fact). To quote from Motz and Weaver:[17]

> "Kepler spent many fruitless hours trying to squeeze Brahe's Martian orbital data into a circle and almost succeeded. At one point, however, the circle he had constructed to fit the data departed from Brahe's observations by 8 minutes of arc. Kepler reluctantly rejected the idea of a circular orbit for Mars, for he knew that Brahe had been too accurate an observer to have made so large an error as 8 minutes of arc! Kepler

then turned to other possible orbits and finally chose the ellipse."

A triumph, on the one hand, of accurate, systematic, and sufficient data, and on the other hand, of an approach to and practice of science whereby theories and hypotheses are formulated precisely and in such a way that they can be tested against accurate observations. The seeds of the *scientific method* were being sown.

A new era began soon after with Galileo Galilei (1564-1642), for with Galileo was launched the development of the *scientific technique* and the emphasis on measurement and ability to measure; the appreciation of the power of the scientific instrument. Galileo developed the telescope, discovered the thermometer, used the motion of the pendulum to measure time and was thus the first to discover that time is a physical quantity, and established the science of kinematics (a description of how objects move). Galileo's studies of mechanics showed that a body does not come to a halt when the force propelling it is ceased, in contrast to Aristotle's view that the continuous motion of a body was possible only if it remained in contact with the applied force. His experiments on the motion of bodies also led him to conclude that it is not the body's velocity,[18] but rather the rate at which the velocity of the body changes with time - - the body's acceleration - - which determines its motion under the influence of gravity. Galileo's kinematics led to Newton's dynamics, to the inquiry into the causes of motion and into the *force* as the cause of the changes in the state of motion of a body.

Isaac Newton (1642-1727) formulated the necessary physical laws explaining the dynamics of the macrocosmic universe and concluded in this way the work of Galileo, Kepler, and the Greeks. He searched for an answer to the question "What is the exact connection between force and motion?" and was led to his three laws of motion described in his book *Principia* (1687) which served as the foundation for the study of dynamics.

Let us focus a little on Newton's second law - - the acceleration of an object is directly proportional to the force acting on it and inversely proportional to its mass - - formally expressed as[19]

$$F = ma. \tag{1}$$

Equation (1) relates the acceleration, *a,* of a body -- in magnitude and direction -- to the force *F* acting on it (the vector quantities *F* and *a* are indicated by bold-face type). The acceleration of a body, it says, equals the force exerted on it times an intrinsic property called the *inertial* mass *m* of the body. Therefore, knowing the force acting on a body and its initial position and velocity, allows complete and precise prediction of the body's position and velocity for all future times. The law represented by Equation (1) is independent of the nature of the force acting on the body. This simple mathematical equation is a masterpiece of magnificent synthesis epitomizing two thousand years of scientific effort. It is one of the simplest, most fundamental, and most general relations in all of physics and in all of science. Known as Newton's equation of motion, is the basis of classical mechanics. It dominated physics until the beginning of the 20th century when it was modified in uniquely remarkable ways by relativity and quantum theory to cover areas of the universe beyond the macrocosmos, the new world of the microcosmic universe where the sizes of objects are exceedingly small and their speeds are exceedingly large.

Newton's synthesis reached a second high when he unified the celestial and the terrestrial worlds through his law of universal gravitation. The force that attracts to earth the objects that are near to the earth (the force that causes the apple to fall from the apple tree!), the force that makes the moon orbit the earth, and the force that causes the earth and the planets to orbit the sun is one and the same: gravity. It is exerted between all bodies in the cosmos. It is universal. Thus, there is nothing special about the attractive force of the earth. The origin of the gravitational force of the universe is the mass *m* of the terrestrial and celestial bodies. And so the law of universal gravitation was born

$$F_g = G\,(m_1 m_2)\,/\,r^2. \tag{2}$$

Two bodies of masses m_1 and m_2 exert an attractive force F_g on each other which is inversely proportional to the square of the distance *r* that separates them. In Equation (2), *G* is a constant of proportionality (the so-called gravitational constant equal to 6.67×10^{-11} m^3 kg^{-1} s^{-2}). Equation (2) is as elegant in its simplicity as is powerful in its predictive capacity. It represents another magnificent unification: massive bodies, anywhere, move with regard to each other because an attractive force F_g is exerted among them. Although, as we shall see later, there are more forces in the universe, the dynamic motions of the macroscopic universe can be attributed to gravity, the prime force in the cosmos.

With Newton began the sharpening of the definitions of the scientific vocabulary, foremost the basic concepts of space and time, and the derived quantities of velocity and acceleration. In his effort to define the instantaneous speed[18] of a particle, Newton invented[20] the differential calculus, one of the most powerful analytical tools developed by man. Space and time, were assumed to be independent of each other, absolute, and infinite in extent. Space geometry was assumed to be Euclidean or flat, distances between events to be the same for all observers, and time to flow continuously from an infinite past to an infinite future. In time, relativity and quantum theory have changed these Euclidean and Newtonian concepts (Section 3.5).

And so the slow journey of science that began in mystery, passed from astrology to astronomy, from the geocentric to the heliocentric description of the solar system, from the circular to the elliptic orbits of the planets, from kinematics to dynamics, and finally to the grand synthesis of Newton and classical mechanics. The protagonists at the end of this long journey had, as those at its beginning, a deep reverence for the Creator of the magnificent cosmos they were so deeply contemplating. However, much had changed, especially during the 100 years from Copernicus to Newton. The experimental method was invented and was being employed as a primary tool for discovery and as a basic criterion of scientific truth. With that the authority of the old masters, the scientists/philosophers of antiquity, was questioned: the properties of the universe could be deduced not solely by logical argument from philosophical principles and *a priori* axioms, but more so from direct observation and experimental-fact-based reflection.

With Newton the slow progress of science ended and the accelerated progress of science began. The starting off platform for this new tempo was Newton's two books: *Principia* and *Opticks*. The former is mathematical, principally describing the laws of motion and the law of gravitation and their deductive consequences. The latter is more experimental, principally dealing with Newton's experiments with light and the formulation of the corpuscular theory of light (light, he believed, is composed of particles, later called photons).

3.3. THE SCIENTIFIC AND THE INDUSTRIAL REVOLUTIONS

Although, as Snow put it, dating the scientific revolution is "a matter of taste,"[21] the middle of the 17th century may well be regarded as its beginning. This period also may be taken as the beginning of the systematic investigation of major areas of the universe in spite of the fact that it was not until the 19th century that the fruit of this revolution ripened as can be seen from Table 3.3 where is listed a sample of the outstanding scientific discoveries which occurred during the latter part of the 19th century.

From the beginning of the scientific revolution there were calls for emphasis on the application of science to the solution of social problems and the practical needs of man. Francis Bacon was an early advocate of this view. Thus, the scientific revolution unavoidably led to the industrial revolution. The latter had arisen from wider human interactions and has had far reaching social consequences. In the 17th century the specific interrelationships between science and technology were minimal, but they increased considerably by the end of the 18th century. For instance, basic investigations in chemistry were triggered in response to needs for bleaching and dying of cloth, and the discovery of urea by Wohler in 1828

Table 3.3. Some outstanding contributions to science between 1850 and 1900

Principal scientist(s)	Scientific discovery/Contribution	Year/Period
L. Foucault/ A.-H.-L. Fizeau	Measurement of the speed of light in water and air	1850
C. Darwin/ A. R. Wallace	Theory of evolution	1859
G. R. Kirchhoff/ R. W. Bunsen	Spectroscopic analysis	1859
F. A. Kekulè	Structural theory of chemical compounds	1858-66
J. C. Maxwell/ R. Clausius	Kinetic theory of gases/ Average kinetic energy of gas molecules	~1859
L. Pasteur	Discovery of germs	1861
J. C. Maxwell	Electromagnetism	1864
G. Mendel	Theory of heredity (basic laws of genetics)	1865
J. C. Maxwell	Distribution of speeds of particles making up a gas	1867
D. I. Mendeléev	Periodic Table of Elements	1869
L. Boltzmann	Statistical theory	1877
H. R. Hertz	Generation and detection of electromagnetic waves	1887-8
W. C. Roentgen	Discovery of X-rays	1895
H. Becquerel	Discovery of radioactivity	1896
J. J. Thomson	Discovery of the electron (determination of the ratio e/m)	1897
M. Curie	Discovery of radium	1898
M. Planck	Quantum theory of radiation	1900
P. Villard	Discovery of γ-rays	1900

opened up the way for the synthesis of medicaments and dyestuffs.[22] Watt's steam engine (around 1765) signifies perhaps the onset of the industrial revolution. Initially, the basis of this revolution was invention and not science, but by the closing of the 19th century the interplay between scientific discovery and industrial innovation began to emerge as can be seen from Table 3. 4 where some technological advances between 1855 and 1904 are listed. Ultimately, this interlocking of basic discovery and technological innovation led to the emergence of the chemical, the engineering, the electricity, the electronics and the transportation industries, as well as the many industrial uses of atomic particles. In this way technologies were established as systematic disciplines to be taught and learned, and science began to reorient progressively a larger part of itself toward feeding the new technologies (Chapter 7).

Table 3.4. Some technological advances between about 1855 and 1904

Technological advancement	Year/Period
Beginning of synthetic drug and dye industry	1856
Improved explosives (smokeless powder, dynamite, etc.)/armored vehicles	1855-67
Industrial dynamo	1860-70
First trans-Atlantic cable	1865
First plastic	1870
Telephone/electric lights/phonograph	1870-1880
Radio/wireless telegraphy	1890-1900
Steam and gas turbines	1896
Powered airplane	1903
Radio tube	1904

3.4. THE FOUNDATIONS OF MODERN SCIENCE

The period from the beginning of the 18th century to the beginning of the 20th century can be considered as the time during which the foundations of modern science were laid. These were the 200 years that led from Newton to Einstein, from the macrocosmos to the microcosmos, from classical to quantum physics. A period of critical observations, ingenious experiments, unique insight, incremental understanding, patient and often independent steps along tortuous paths that in time converged and led to brilliant syntheses and bold propositions. A time of gradual step-by-step steady evolution, occasionally interrupted by revolutionary discoveries, and steep step-function-type (abrupt) advances. In this time period many of the fundamental fields of modern science were developed in parallel or in tandem cross-fertilizing and cross-breading. Discovery bread discovery, innovation superseded innovation, and in a chain-reaction-like process some of the broadest laws of science were established. We shall follow parts of this trail highlighting some of the major accomplishments in the physical science. Breathless we would have remained if we attempted to follow the trail in all its tortuous and majestic intertwining loops. But even a selective portion of this path - - that which led to the great discoveries in physics of the late 19th and early 20th century and to the galloping scientific advances beyond -- will help us appreciate the way science progresses and the way science evolves. By the end of this time period all prerequisites for the transition from classical to quantum mechanics, from the macrocosmic to the microcosmic universe, were essentially in place. Four fundamental constants dominated physics at the end of this period: the electron charge, e; the quantum of action known as the Planck constant, h; the gas constant per molecule known as the Boltzmann constant, k; and the speed of light in vacuum, c.

3.4.1. Development of the Mathematical Foundations of Modern Science

Following Newton, classical physics and celestial mechanics were further developed by such towering figures as P. S. Laplace, J. L. Lagrange, J. B. Fourier, W. R. Hamilton, S.-D. Poisson, C. G. J. Jacobi, and H. Poincaré. Powerful new mathematical methods such as differential calculus and partial differential equations, and concepts such as the potential energy and the Lagrangean function were constructed which formed the foundations of modern science and accelerated the development of the then emerging fields of electricity, magnetism, heat, and the behavior of large number of particles in thermodynamic equilibrium. By the end of this period these new fields grew to maturity. Their broad laws and general principles while modified later by relativity and quantum mechanics remained the same since their inception during this period.

3.4.2. Electricity, Magnetism, and Electromagnetism

The ability of some bodies to attract light objects (straws or bits of paper) when rubbed with cloth was known in Greek antiquity. In spite of this knowledge, it was not until the 17th century that the study of static electricity began. The phenomenology of electrostatics and magnetostatics was largely established in the 18th century.[23] A multitude of step-by-step discoveries (Table 3.5) led to the development of the fields of electricity and magnetism, and ultimately electromagnetism; new fields not reducible to mechanics. First, it was the advances in static electrification and electrostatic machines. Then, it was the discovery of conductors and insulators, the existence of two - - and only two - - types of electricity (the positive and negative charges of today), the conservation of electricity (the algebraic sum of positive and negative charges is constant), the repulsion between like charges and the attraction between opposite charges, the electrical capacitor, and electrical induction. By about 1770 sufficient step-by-step progress had been made and sufficient basic facts had been

Table 3.5. Some advances in electricity, magnetism, and electromagnetism in the 18th and 19th centuries[a]

Principal scientist(s)	Scientific discovery	Year
S. Gray	Discovery of conductors and insulators	< 1736
C. Dufay	Discovery of the two (positive and negative) kinds of electricity, charges[b]	~ 1733
E. G. von Kleist	Electrical condenser (capacitor)	1745
J. Canton	Electrical induction	1753
C.-A. de Coulomb	Coulomb's law of force for electric charges	1788
L. Galvani	Discovery of electric current	1791
A. Volta	Discovery of Voltaic cell	~1800
H. C. Oersted	An electric current generates magnetism/ First connection between electricity and magnetism	1820
M. Faraday	Electromagnetic induction	1831
M. Faraday	Laws of electrolysis	1833
L. Foucault/ A.-H.-L. Fizeau	Measurement of the speed of light in air and water	1850
J. C. Maxwell	Electromagnetism	1864
H. R. Hertz	Experimental demonstration of electromagnetic waves	1887

[a] Based, in part, on Refs. 10 and 25.
[b] It took about 100 years to establish the existence of positive (+) and negative (-) charges.

discovered to make inevitable a unified understanding of the observed phenomena through a quantitative law. Indeed, in 1788 Charles-Augustin de Coulomb (1736-1806) formulated his inverse square law of force for electric charges (reminiscent of Newton's law for the mass)

$$F = k\,(q_1 q_2 / r^2).\tag{3}$$

Two (point) charges q_1 and q_2, at a distance r, exert on each other a force F along r which is inversely proportional to the square of the distance r (k is a proportionality constant). Coulomb's law is the basis of electrostatics and the basis of magnetostatics for magnetic monopoles.[24] It describes the force between two charges when they are at rest. Additional forces come into play when the charges are in motion.

Even though many of the electric phenomena during this period implied the flow of electric charges (electric currents), the idea of an electric current in the sense of a continuous flow of charge that could be set up and controlled at will was not yet born. Such a current was discovered accidentally in the 1780's by Luigi Galvani (1737-1798) who observed that the muscles of a frog's legs in contact with two unlike metals (zinc and copper) contracted spasmodically. Galvani's suggestion that the phenomenon was biological in nature was rejected by his colleague Alessandro Volta (1745-1827) who proposed that the response of the frog was caused by an electric current generated by the two different metals in contact with the nerve of the frog's leg, i.e., the frog's nerve acted as a wire, as the conductor of the current. Volta then showed that if the two dry ends of a copper and a zinc rod were immersed in sulfuric acid and were connected by a wire, a current flowed through the system. This was the birth of the Voltaic cell, the first battery. A discovery that profoundly impacted society. The process that led to this discovery is a characteristic example of how science progresses step-by-step: an accidental phenomenon took place , a skillful eye observed it, a knowledgeable expert from another field conjectured an explanation, experiments were designed to test the proposition, the discovery of the electrical current was confirmed, and a new device was invented that provided a new tool to further interrogate nature and to serve society.

In Volta's discovery lay the beginnings of the electrical industry and the development of instruments to measure current and voltage. With it electrical research shifted rapidly from electrostatics and the study of phenomena using static charges, to moving charges and the effects of passage of electric current through various substances, electrolysis, and inevitably the discovery of electromagnetism. Indeed, in 1820 Christian Oersted found accidentally that *an electric current generates magnetism.* The discovery of the connection between electricity and magnetism was followed by Michael Faraday's discovery in 1831 of electromagnetic induction. If an electric current could produce a magnetic filed a changing magnetic field had to be able to produce an electric current. The generation of an electric current in a conductor which cuts the lines of the magnetic force of a moving magnet led to the development of the electrical generator or dynamo which transforms mechanical energy to electric. The two symmetrical discoveries of Oersted and Faraday - - production of an electric current by the motion of a magnet (Faraday) and the influence on a magnet by the motion of electric charge (Oersted) - - demonstrated that electricity and magnetism are connected. They signaled the moment that the step-by-step discovery in the separate fields of electricity and magnetism finally reached a point when the parts could be integrated into a coherent whole, electromagnetism.

The ultimate unification of electricity and magnetism was accomplished through the work of another great synthesizer, James Clerk Maxwell (1831-1879). In Maxwell - - as earlier in Newton and later in Einstein - - ingenuity, mathematical skill, and a unique ability to synthesize and unify the seemingly different and envision the existing within a new conceptual frame, terminated the step-by-step advancement in a particular field of science and abruptly elevated the understanding via a new paradigm to a higher plateau wherefrom the-step-by-step progress began anew. A prerequisite of such syntheses always was an accumulation of reliable scientific facts and ideas, and anomalies and inconsistencies in the rationalization of existing facts and observations. Indeed, Maxwell's synthesis became possible because of a large body of work on and understanding of the nature of electricity and magnetism that preceded it.

Besides the key discoveries referred to above, Maxwell stood on a platform of knowledge that was supported by Faraday's concept of the field, Gauss's law of electricity, Gauss's law of magnetism, and Ampère's law relating the current flowing in a wire to the magnetic field around it. In his effort to explain an inconsistency regarding the electric field associated with a parallel-plate capacitor whose plates are connected by a conductor, Maxwell introduced the concept of the displacement current, and by working out the electric oscillations of the capacitor, he was led to the electromagnetic oscillator and the discovery of the wave character of electromagnetism. A changing electric field, he showed, generates in empty space a magnetic field -- and is, in essence, equivalent to an electric current. Furthermore, if a changing magnetic field produces an electric field, the electric field will itself be changing. In turn, this changing electric field will produce a magnetic field which will be changing and thus will produce a changing electric field, and so on. The net result of these interacting changing fields was a wave of electric and magnetic fields - - an electromagnetic wave - - that can propagate through space. The variations of either field (electric, E, or magnetic, B) from point-to-point in space with the variation of the other field (magnetic or electric) with time are related through a set of elegant equations bearing Maxwell's name summing up his historic synthesis. In the absence of dielectric or magnetic materials, Maxwell's equations can be written as:[26]

$$\oint E.\ dA = Q/\varepsilon_0 \tag{4a}$$

$$\oint B.\ dA = 0 \tag{4b}$$

$$\oint E.\ dl = -\,d\Phi_B/dt \tag{4c}$$

$$\oint B.\ dl = \mu_0 I + (\mu_0 \varepsilon_0)\,d\Phi_E/dt. \tag{4d}$$

They encompass all previous laws of electricity and magnetism. Equation (4a) is a generalized form of Coulomb's law relating the electric field to its source, the electric charges Q; it is Gauss's law of electricity. Equation (4b) represents the same for the magnetic field except that since there are no magnetic monopoles, the magnetic field lines are continuous; it is Gauss's law of magnetism. Equation (4c) states that an electric field E is produced by a changing magnetic field of flux Φ_B; it is Faraday's law of induction. Equation (4d) says that a magnetic field B is produced by an electric current I or a changing electric field of flux Φ_E; it is Ampere's law as modified by Maxwell.

Equations (4a - 4d) are basic for all electromagnetism and are valid even relativistically. They contain two experimentally-determined fundamental constants from electrostatics and magnetostatics: the permeability of free space, μ_0, specifying the strength of the magnetic force, and the electric permittivity of free space, ε_0, specifying the strength of the electric force. These two constants were shown to determine another constant, the speed, c, at which the electromagnetic signals travel in free space: $c = 1/(\mu_0\varepsilon_0)^{1/2}$. Thus, *accelerating electric charges generate electromagnetic waves traveling with the speed c independently of their wavelengths: all electromagnetic waves have the same speed when traveling in vacuum.* Maxwell's synthesis of electricity and magnetism has, thus, become a theory of light. The speed of light as it appears for the first time in Maxwell's equations has become one of the most fundamental constants in all science, central to the theory of special relativity that followed four decades later. The way toward understanding the nature of light had become wide open. The existence of electromagnetic waves themselves was demonstrated in 1887 by Heinrich Rudolf Hertz, and the relationship between light and electromagnetism was further substantiated by the later discoveries of the effects of magnetic and electric fields on light, respectively known as the Zeeman and the Kerr effects. As to the impact of these developments on society, the generation and transmission of electromagnetic waves and radio communication have changed the world and triggered a technological revolution the end of which is still not in sight.

3.4.3. Kinetic Theory of Gases

By the beginning of the 19th century, air was recognized as having weight, atmosphere as exerting pressure, and models of a gas consisting of freely moving particles as significant in understanding applied problems, for instance, in hydrodynamics. By the beginning of the 19th century also the ideal gas law and the equation of state for an ideal gas (one whose atoms or molecules do not interact with each other) were established. The equation of state for an ideal gas

$$PV = nRT \tag{5}$$

represents the end result of another step-by-step evolutionary process and unifies the findings, and rationalizations of more than 200 years of effort (Table 3.6). It relates the macroscopic quantities of gas pressure P, and gas temperature T -- which are used to describe the macroscopic behavior of the gas -- to the large number of microcosmic particles, the atoms or molecules making up the gas. In Equation (5), V is the gas volume, n is the number of gas moles, and the symbol R is the gas constant which is equal to 8.317 joule/(g-mole) K. One mole, or one gram molecular weight, is an amount of gas whose mass in grams is equal to its molecular weight. The number of molecules in one gram-mole of gas is called the Avogadro number and is denoted by N_A. Although Avogadro had no estimate as to how large the number N_A was, it is a very large number indeed, 6.025×10^{23}. In Equation (5) the number of molecules per unit volume (cm^3), N, is related to n by $N = nN_A /2500$. The equation of state, thus, relates the macroscopic and microscopic properties of gaseous matter and shows that the former can be understood from the latter. For instance, the gas temperature is merely a measure of the average kinetic energy of the molecules making up the ideal gas. It was indeed shown by Maxwell in 1859 that the average kinetic energy of an ideal-gas molecule at a temperature T is equal to $3/2\ kT$, where $k\ (= R/N_A)$ is the Boltzmann constant.

The synthesis regarding the behavior of large numbers of particles in thermodynamic equilibrium comprising gaseous matter continued with the work of Rudolf Clausius, James Clerk Maxwell, and Ludwig Boltzmann. Thus, Clausius in 1858 introduced the concept of the mean free path (the

Table 3.6. Some advances in kinetic theory of gases and statistical mechanics

Scientist	Discovery	Year
R. Boyle	Ideal gas law	1662[a]
J. A. C. Charles	Ideal gas law	1787[a]
J. L. Gay-Lussac	Ideal gas law	1802[a]
A.-L. Lavoisier	Conservation of mass in chemical reactions/ Distinction between elements and compounds	<1800
J. Dalton	Experimental basis for atomic constitution of matter	1808
A. Avogadro	Avogadro's number/ Equation of state for ideal gas	1811
R. Clausius	Concept of mean free path	1858
J. C. Maxwell	Average kinetic energy of molecules in a gas/ Distribution of speeds of molecules in a gas	1859
R. Clausius	Concept of entropy	1865
J. D. Van der Waals	Equation for imperfect gases	1873
L. Boltzmann	Statistical theory	1877

[a] Ref. 27.

average distance a molecule travels before it collides with another molecule), and Maxwell in 1859 formulated the description of the distribution of speeds of the particles making up a gas and its relation to the nature of the gas and the gas temperature and described the equality of the mean kinetic energy among the gas molecules. Similarly, in 1877 Boltzmann showed how statistical theory can give rise to the irreversible behavior of macroscopic physical quantities when these are defined in a statistical way, as the physical properties of large complicated systems inevitably are. The mechanics of the kinetic theory of gases were being shown to be part of statistical mechanics.

Important as the ideal gas law is, it applies to an ideal gas and is thus limited. The behavior of gases especially at high pressures and low temperatures was found to diverge from Equation (5). They, for instance, liquify. This led in the late 18th and the 19th century to the study of imperfect gases and a consideration of the forces between the atoms and the molecules of a gas, as well as their finite volumes. In 1873 Van der Waals derived an equation for an imperfect gas which was followed by detailed descriptions of the various states of aggregation of matter.

Today, the four states of matter -- gas, liquid, solid, and plasma -- are studied by powerful experimental techniques and sophisticated theoretical formalisms and their properties are manipulated by intricate ways involving finely tuned energy exchanges between radiation and matter. Even the density regions between the gaseous and the condensed phases of matter are being bridged by interphase (high-pressure) and clusters studies.[28] The latter- - clusters - - are aggregates of atoms or molecules generally intermediate in size between individual atoms, and aggregates large enough to be called bulk matter. One type of clusters, the Van der Waals clusters, are based on the Van der Walls equation for an imperfect gas.

The end of the 18th century may be taken as the dawn of chemistry as a scientific discipline. With it came the first evidence for the atomic constitution of matter. By this time a sound basis was built (largely through the work of Antoine-Laurent Lavoisier) for the discovery of the conservation of mass in chemical reactions and the distinction between elements and compounds, the latter being a combination of two or more elements. This had led to two quantitative regularities of general nature by John Dalton which enabled him to propose in 1808 the atomic theory of matter. The first regularity was that the proportions by weight of two elements in a chemical compound are always the same no matter how the compound is produced (e.g., in water, oxygen and hydrogen are always in the ratio 8:1). Dalton then recognized that the ways in which elements combine to form the various compounds (the non-elemental substances, the molecules), could be understood if each element was an indivisible entity -- an *atom*, the atom of Democritus -- in any chemical reaction. Molecules contain one or more atoms of each kind in definite ratios. The second regularity is referred to as the law of multiple proportions according to which when two elements form more than one molecule with each other, the proportions by mass of the two elements in the different molecules are simple fractions. From the weight (mass) relations of atoms in molecules a scale of relative weights of atoms was subsequently established.

The discovery of the atomic nature of matter[29] accelerated the study and understanding of chemical phenomena and led to the realization that there is a high degree of order in the chemical behavior of different elements. This culminated in 1869 in the Periodic Table of Elements when Dmitri Mendeléev discovered that by arranging the chemical elements in order of increasing atomic weight from hydrogen (the lightest atom) up to uranium (the heaviest known element at the time) caused elements with similar properties to recur at regular intervals. Different elements were assigned positions in a rectangular table, vertical columns of elements, the *groups,* and horizontal rows of elements, the *periods,* with the atomic weight increasing in going down a column and to the right along a row. Elements in a particular vertical group possessed similar chemical properties. The periodic variation in the chemical and physical behavior of the atoms as their weight increases is significant to this day, although the original simple table has since become complicated for reasons which are now well-understood.[30] The periodicity in the chemical and physical behavior of the elements, as well as the amazing stability of the elements, became apparent later when the electronic structure of the atom was established by quantum mechanics during the early part of the 20th century.[31]

Facts and empiricism once again preceded deductive generalization. The Periodic Table of the Elements since its inception provided a basis for understanding the chemistry of the more than 100 elements and their compounds as well as the trends in their chemical and physical properties. It stimulated new developments in science, especially in chemistry. But perhaps the most fascinating "element" of all the elements is the realization that *everything* in the cosmos is made up of these same and limited-in-number elements and nothing else. The infinite variety of nature lies in the infinite combination of these elements and their recycling. Yet, by the beginning of the 20th century it was becoming increasingly clear that atoms were not the most fundamental (elemental) entities in nature. The sequence of events that led to the unraveling of the inner structure of the atom is a distinct accomplishment of man.

3.4.4. Thermodynamics and Statistical Mechanics

These two fields of science are related to the kinetic theory of gases in that they, too, are concerned with the behavior of systems composed of a large number of particles. Through them the ideas of conservation of energy, equivalence of heat and mechanical energy, entropy, and the statistical interpretation of the macroscopic properties of matter entered the field of science.

As we have briefly discussed in Chapter 2 when we were discussing the subject of energy, thermodynamics rests on the law of conservation of energy (the first law of thermodynamics) and the law of entropy (the second law of thermodynamics). These are broad laws which are valid under all conditions we know of and are applicable to all forms of matter and energy and their interactions (absorption, emission, and scattering of energy by matter). Both laws impacted the development of science and science's benefit to society. Let us, then, focus a little on the second law of thermodynamics which deals with the concept of *entropy, S,* introduced by Rudolf Clausius in 1865. He defined the ratio of the amount of heat Q transferred to a body at a temperature T to that temperature, to be the change, ΔS, in the entropy of that body, i.e.,

$$\Delta S = Q/T. \tag{6}$$

The entropy of an isolated system is not conserved (as does its energy), but it always increases with time until it reaches a maximum at equilibrium. The concept of entropy is thus connected with time for it introduces a distinction between the past and the future: the past is the time when the total entropy of an isolated system was smaller than is now. The explanation of this macroscopic asymmetry in the world by the low-entropy initial state of the universe makes it necessary to hypothesize that in the past the universe was more ordered than it is today.[32-35] The second law of thermodynamics then allows us to define what has been called "the arrow of time": the past and the future look different for there was less entropy in the past and there will be more entropy in the future. The second law of thermodynamics also allows us to distinguish reversible from irreversible processes. An example of a reversible process is the perfect elastic collision between two spheres as is considered in mechanics and an example of an irreversible process is the expansion of a gas in vacuum. Since the arrow of time points in the direction of increasing entropy, increasing disorder, and since only irreversible processes contribute to the production of entropy, the direction of natural processes is that of irreversibility. Natural phenomena are irreversible - - none of us has ever seen a gas spontaneously compress itself, or a bucket of water poured into the ocean spontaneously be collected back into the bucket (such events are improbable). This distinguishes processes involving heat from those of mechanics and electromagnetism. The latter two are reversible in time, the former is not. In classical and quantum mechanics the fundamental laws of physics are taken to be symmetrical in time.

If the macroscopic phenomena are a combination of elementary microscopic phenomena which are themselves *reversible* how can the *irreversibility* of the macroscopic world be understood? The search for an answer to this profound 19th-century question led to the emergence of a new field of science, statistical mechanics, which deals with systems comprised of many particles (systems with many degrees of freedom). It is the field of science that introduced -- foremost through the work of Maxwell, Clausius, Boltzmann, and Gibbs -- the concept of probability into physics. The entropy of a system had now been related to the statistical behavior of the particles which make up the system. Consider, for instance, an atomic gas at atmospheric pressure. In every cubic centimeter of space the gas occupies there are more than 10^{19} atoms. A statistical analysis of such a system by Ludwig Boltzmann (1844-1906) made a distinction between the "macrostate" of the gas and the "microstates" of the gas. The macrostate is specified by the macroscopic properties of the gas such temperature, pressure, and entropy. The microstate of the gas would be specified if the position and the velocity of each and every gas atom were known. There is no way of knowing the latter for such a huge number of particles. Only the macrostate of the system can be known. However, for every macrostate there are many microstates, many different ways in which the positions and velocities of the atoms of the gas can combine. The number of these combinations constitutes the number of microstates, W, of the gas and determines the probability of the gas being in a given macrostate. The larger the number W of microstates that correspond to a given macrostate, the larger the probability that the system will be in that macrostate. That is, the probability of a given macrostate is directly related to the disorder of the system. With this reasoning, Boltzmann showed that the entropy, S, of a given macrostate is related to W by $S = k \log W$. The irreversible macroscopic quantity entropy is thus interpreted by statistical theory via the reversible microscopic processes embodied in W. Entropy is a thermodynamic measure of the system's disorder and a measure of the probability of the state of the system. Both, probability and disorder, tend to increase with time in an isolated system. And yet, the question "How does the macroscopic thermodynamic description of a system relates to the microscopic description of the particles it is comprised of?" proved difficult to answer as the arguments between Poincaré and Boltzmann and the continued scientific and philosophical discussions since[36] would attest. The individual gas molecules do not follow the second law of thermodynamics. Thermodynamics and dynamics are incompatible concluded Poincaré. *There is a complementarity between the microscopic dynamical microstate-type description and the macroscopic macrostate-type description.*

Out of statistical mechanics, came, in 1859, Maxwell's law describing the velocity distribution of the particles making up a gas. The number of molecules in a given velocity interval is proportional to the density of the gas and depends only on the gas temperature and the mass of the particles the gas is comprised of. The law had been experimentally confirmed by O. Stern in 1921. Out of statistical mechanics also came predictions which were difficult to rationalize on the basis of the then existing experimental facts. For instance, the prediction that any degree of freedom of a system in thermal equilibrium has the average kinetic energy $\frac{1}{2} (kT)$ implied -- contrary to reality -- that the internal degrees of freedom of an atom did not contribute to its specific heat.[37] Troubling inconsistencies, the ingredients for the next level of evolution in man's understanding of nature.

3.5. A NEW FRAME

By the closing years of the 19th century, classical mechanics was found incapable of explaining a number of important new phenomena. Distinct among these was the propagation of electromagnetic waves through empty space. In the view of the then leading scientists this required a medium upon which the electromagnetic waves traveled, but mechanical models assuming the existence of such a medium (called "aether") were unsatisfactory. Newtonian mechanics required the speed of light to be different for observers moving in different directions through aether, in contrast to the prediction of Maxwell's equations that the speed of light is the same regardless of the state of motion of the observer. According to Newtonian mechanics also an object can, in principle, be accelerated to a speed greater than the speed of light if a force acts on it sufficiently long, which is in contradiction with Maxwell's electromagnetism according to which a charged particle cannot move faster than the speed of light. The first observations of energy exchanges between radiation and matter: the emission and absorption line spectra of chemical elements, the weak ionization of gases produced by ultraviolet (UV) light, the photoelectric effect, and the glowing of hot objects were similarly unaccounted for. Classical physics seemed incompatible with these new phenomena. It was *time for a new synthesis and a new paradigm; time for a revision and a new vision.* This came from two masters -- Max Planck and Albert Einstein -- with deep knowledge of thermodynamics and statistical mechanics and enormous scientific intuition. The resolution of the new phenomena regarding the emission and absorption of electromagnetic radiation by matter led to quantum physics and to light quanta. The resolution of the propagation of electromagnetic waves in empty space led to the special theory of relativity and the revision of the concepts of space and time. In both cases the experimental facts preceded and led to a successful theory.

3.5.1. The Quantization of Energy and the Quantum Theory of Light

Let us first follow the path of light and briefly summarize the knowledge that existed at the end of the 19th century regarding the origin and nature of light. The trail of light began with Newton and is still today full of adventure and amazement. Newton considered light to be corpuscular in nature, but before this came to be recognized 200 years later, the ingenious work of many physicists -- foremost Christiaan Huygens (1629-1695), Thomas Young (1773-1829), and Augustin Jean Fresnel (1788-1827) -- established the wave nature of light.

- Light is reflected, refracted (when a wave traveling in one medium crosses a boundary into another medium where its velocity is different, the transmitted wave may move in a different direction than the incident wave), and diffracted (light bends around encountered obstacles and passes into the region behind them; the amount of diffraction depends on the wavelength of the light and the size of the obstacle).

- Light waves show interference (they reinforce or cancel each other when they pass through the same region of space at the same time), are transverse (the direction of motion of their vibrations is perpendicular to the direction of their propagation, not longitudinal as in sound), can be polarized.

- White light can be separated into colors and the wavelength of different color light can be measured.

- Monochromatic light can be coherent, that is, there is a predictable correlation between the amplitude and phase at any point on the wave and at any other point.

A little later, in 1850, the speed of light in air and in water was measured by Léon Foucault and Armand-Hippolyte-Louis Fizeau and was found to be greater in air because the index of refraction of air is less than the index of refraction of water, in support of the wave nature of light.

As we have seen earlier in this chapter, the trail of light was illuminated by advances in two other fields: electricity and magnetism. Maxwell proposed that light is an electromagnetic wave traveling at the speed c and Hertz demonstrated the electromagnetic nature of light: electromagnetic waves produced in an electrical oscillator propagated as Maxwell's equations predicted. They were reflected, diffracted, and refracted exactly as light is. Other experimental investigations observed light emission from hot solids, sparks, liquids, and gases, implying that electrical charges are oscillating in matter when matter emits light. Thus, matter must consist of electrically charged particles although matter was known to be neutral.

Spectroscopy was another source of new phenomena and new questions. It began with Joseph von Fraunhofer's observation around 1815 of a pair of bright yellow lines in the spectrum of a sodium lamp and continued with the work of Robert Wilhelm Bunsen (1811-1899) and Gustav Robert Kirchhoff (1824-1887) who around 1860 identified these lines as signals from atoms and molecules which were later associated with the electronic and molecular structure and served as atomic and molecular "fingerprints." Although at the time this explanation could not be provided, these observations were the beginnings of the field of Analytical Chemistry. They clearly demanded an answer to the question: what is the electromagnetic object that could emit the observed spectra? Maxwell's theory could be used to calculate the radiation by moving charges, but it could not be used to explain the observed atomic and molecular radiation. The light of the trail of light was still dim. Classical theory could explain neither these light emissions nor the careful and systematic measurements on blackbody radiation conducted at the Physikalisch Technische Reichsanstalt in Berlin. Efforts to understand the latter led to the concept of quantization of energy.

Blackbody or thermal radiation is light emitted by a body as a result of its temperature. It is of universal character in that it depends only on the blackbody temperature and not on its composition. The spectrum of the emitted light is continuous, that is, it contains electromagnetic waves of all frequencies, ν, but each frequency has different intensity. The energy emitted per unit time in the frequency range $\nu + d\nu$ from a unit area of the blackbody at absolute temperature T is known as the spectral radiancy $\rho_T(\nu)d\nu$. Reliable and detailed measurements of this quantity had been made as a function of the frequency ν at various blackbody temperatures which could not be fully explained on the basis of existing theory in spite of the persistent efforts of prominent scientists such as G. R. Kirchhoff, W. Wien, J. Stefan, Lord Rayleigh, and J. H. Jeans. Their work gave rise to laws bearing their names, but these laws were at best partially successful. For example, Rayleigh's law $\rho_T(\nu)d\nu = (8\pi\nu^2 d\nu kT)/c^3 d\nu \propto \nu^2$ while successful at long wavelengths leads to the unphysical prediction that the total emissive power integrated over all frequencies is infinite, in complete disagreement with the experimental facts.

The complete understanding of the experimental facts came from the work of Max Planck (1858-1947) and required the introduction of a new paradigm. Let us follow this process because it is another classic example of the intricate ways in which science works and evolves. Planck began working on the problem of blackbody radiation in 1897 most likely because it was one of the most important and challenging scientific problems of that time. He first derived by a rigorous combination of electrodynamics and thermodynamics what others did before. He attempted to theoretically simplify the problem and in so doing he strictly relied on what was reliably known. One basic fact which he accepted was that blackbody radiation does not depend on the nature of the walls of the emitting body, but only on its temperature. Planck then thought to analyze the radiation emitted by a blackbody with walls made of Hertzian oscillators whose behavior he could calculate. In this way he simplified the problem by avoiding dealing with the detailed constitution of the atoms or the molecules making up the blackbody's walls, which was not known at the time. With this important simplification, Planck found that the emissive power of the blackbody had to be proportional to the average energy of the oscillators. The oscillators, being in thermal equilibrium, are required by statistical mechanics -- as Maxwell and Boltzmann had shown -- to each have the *same* average energy $\langle\varepsilon\rangle$ proportional to the absolute temperature T, namely, $\langle\varepsilon\rangle = kT$.

The combination of Planck's results on the relation between the average energy of the oscillators and emissive power gave the expression derived earlier by Rayleigh which did not agree with the experimental data except at low frequencies v.

Guided by the experimental data, Planck considered the possibility that the classical law of equipartition of energy may be violated and may, in fact, be the source of the problem. Thus, since theory and experiment generally agreed at low v, he *assumed* that $<\varepsilon> \rightarrow kT$ (for $v \rightarrow 0$), and the discrepancy at higher energy could be eliminated if for some reason $<\varepsilon> \rightarrow 0$ as $v \rightarrow \infty$, that is, if the energy ε of each oscillator was not constant but a function of its frequency v. Planck realized that he could satisfy this latter condition by treating ε as if it were a *discrete* variable instead of a continuous variable as is implied by classical physics. He, then, *assumed* that the energy of each oscillator could only take certain discrete values and finally took $\Delta\varepsilon \propto v$, or $\Delta\varepsilon = hv$, or simply $\varepsilon = nhv$, where n is an integral number. The harmonic oscillator cannot have any energy as classical mechanics taught, but only discrete values, integral multiples of hv. Radiation is emitted only in discrete units of energy hv. With this assumption Planck showed in 1900 that the spectral radiancy of a blackbody is represented by the expression

$$\rho_T(v)dv = (8\pi v^2/c^3)[hv/(e^{hv/kT} - 1)]dv. \tag{7}$$

The experimental results were found to be in perfect agreement with Planck's formula (Equation 7). At low energies ($hv/kT \ll 1$) the new law gives as an approximate expression the classical limit found by Rayleigh and at high energies ($hv/kT \gg 1$) it gives the formula obtained earlier by Wien. A more general law was found, which supersedes the old laws and reduces to them when applied to their restricted regions of validity. However, the new law goes further: it explains the observed behavior of nature for in-between energies where the old laws had failed. The new formula explains the facts, incorporates the predictions of old theories as special cases and reduces to them when applied to the energy regions of their respective domain, extends the range of applicability of the physical law, and has revolutionary conceptual and practical consequences. Yet, at the time Planck announced his discovery, he himself felt that the formula needed theoretical justification. Why did nature seem to obey this idea of quantization? At the time of his discovery Planck himself did not have an answer. Indeed, when in 1931 Planck was asked by the American physicist R. W. Wood[38] how he had invented something as incredible as the quantum theory he said the idea came to him out of desperation after everything else he tried had failed. "It was an act of desperation. For six years I had struggled with the blackbody theory. I knew the problem was fundamental and I knew the answer. I had to find a theoretical explanation at any cost, except for the inviolability of the two laws of thermodynamics."[38]

And yet the idea, the postulate, that the energy of the electromagnetic standing waves, oscillating sinusoidally in time, is a discrete rather than a continuous quantity, was one of the most profound ideas of modern science. A bold assumption was made, and an idea was postulated by a "desperate" master because the experimental method unraveled the true behavior of nature with regard to light emission from hot bodies and because the experimentally revealed behavior of nature remained unexplainable within the frame of the conceptual scheme of the then accepted theory. The experimental data forced the scientist to use his intuition and to go beyond the existing conceptual frame. Faced with the difficulty of unexplained experimental facts, science emerged with a new paradigm: *the quantum of energy. Light is emitted in distinct microscopic "packets," "quanta" of energy.*

The concept of quantization was quickly embraced by Albert Einstein (1879-1955) and later by Niels Bohr (1885-1962) and others to explain the new phenomena of the emerging microcosmos. Planck's initial postulate that the energy of an oscillating particle comes in small units -- is quantized -- was subsequently broadened to include oscillating electromagnetic waves and any quantity whose single coordinate oscillates sinusoidally with time. It led to the quantum theory of radiation. The crucial step from Planck's idea of energy quantization to the quantum theory of radiation was taken in 1905 by Einstein in his traditional approach: he began from an *accepted fact* -- the idea of energy quantization -- *and* explored its consequences. The result was another new paradigm: *the concept of light particles.*

If the energy of the oscillators is quantized, Einstein argued, when light is emitted by them their energy (say, $nh\nu$) must decrease by at least an amount $h\nu$ to $(n - 1)h\nu$, and conservation of energy would require light to be emitted in quanta each of energy equal to $h\nu$. And he went further. He *assumed* that light is not just generated in packets of energy initially localized in a small volume of space, but light remains localized in these packets -- particles named "*photons*" (from the Greek word φῶς, meaning light) -- as it travels through space away from the source with a velocity c. The energy E of each photon is related to its frequency v by $E = hv$.

The new construct about the nature of light explained the existing experimental facts on the emission of electrons when light is shined on metal surfaces, which remained unexplained by the classical theory of light that assumed light to be a wave. If in the photoelectric effect process, one photon gives up all of its energy to an electron in the photocathode material, and if the photon energy exceeds the work function, W, of the photocathode material, the liberated photoelectron must have a maximum kinetic energy $E_{kin} = hv - W$. The experimental results obeyed fully the new theory. Thus, the photonic theory of light was confirmed by its ability to successfully account for the then existing unexplained experimental facts.

And so the path of light in the step-by-step-process of discovery revealed a little more about its nature: light is not just an electromagnetic wave with frequency v and wavelength λ, but light is *also* a particle with an energy $E = hv$, and a momentum $p = E/c = hv/c = h/\lambda$. Light is not just a wave that is reflected, refracted, and diffracted, but *also* a particle which is exchanged or scattered in collisions with matter. It is possible for a photon to give up only part of its energy to a charged particle and be scattered just like a particle. This was demonstrated less than two decades later in 1922 by Arthur H. Compton who showed that photons can be scattered by atomic electrons, the latter been set free just as they are when particles collide with atoms. This is another demonstration of the particle-like aspects of electromagnetic radiation because not only can a precise energy hv be assigned a photon but also a precise momentum h/λ. Both the photon energy and the photon momentum are conserved when photons collide with particles.

Thus, light phenomena can sometimes be understood by considering light as a wave and sometimes by considering light as a particle. The roots for the next level of unification -- that of the concepts of waves and particles -- began to appear. A complementarity (Section 3.6) between the two descriptions of light as a wave and as a particle.

3.5.2. Relativity

Relativity is the natural conclusion[39] of the work of Michael Faraday, James Clerk Maxwell, Henri Poincaré, George Francis FitzGerald, and Hendrik Antoon Lorentz. As discussed earlier in this section, Maxwell's electromagnetic theory of light required that electromagnetic waves travel in empty space with the speed c. All other known wave motions involved the propagation of fluctuations in a definite medium. Thus, it seemed necessary that electromagnetic waves represent disturbances in some medium. A medium was therefore mathematically constructed that was called aether which was supposed to pervade all matter while being itself immaterial. FitzGerald, Poincaré and Lorentz assumed that there is a mechanical aether, that light behaves according to Newtonian mechanics, and that measuring sticks and other bodies moving through the aether change their lengths in such a way that the speed of light appears to be constant. In particular, Lorentz had discovered a coordinate transformation which leaves Maxwell's equations invariant. The Lorentz transformation is expressed by the equations

$$x' = (x - \upsilon t)/[1 - (\upsilon/c)^2]^{1/2} \tag{8}$$

$$t' = (t - \upsilon x/c^2)/[1 - (\upsilon/c)^2]^{1/2}, \tag{9}$$

where x is the space coordinate and t is the time in the first system, and x' and t' are the corresponding quantities in the second system which moves at a velocity υ with respect to the first system. If, however, aether were to fulfil the role of an absolutely fixed reference system,[40] it was necessary that some experimental means be found for measuring the speed of the earth relative to aether. In 1887, Albert Michelson and Edward Morley conducted an experiment to determine whether the time taken for light to pass over two equal paths at right angles was indeed different as it should be if the earth were moving relative to the aether through which the light traveled. No discernible effect due to the earth's motion through the hypothetical aether was found. Hence the earth is perfectly at rest in the aether, a result inconsistent with the existence of the aether. Once again an important problem remained unresolved. Once again the existing theory could not account for the experimental facts. Once again out of a difficulty a new paradigm and a new frontier would emerge. This was done by Einstein in his characteristic approach: instead of arguing whether the speed of light is constant or not, he *accepted the prediction of Maxwell's equations that the speed of light is always the same regardless of one's state of motion and look into the consequences of the assumption.* The result was the special theory of relativity and through it the conceptual unification of electromagnetism and mechanics. He axiomatically postulated that it is not possible to distinguish one reference system from another moving with a velocity constant in magnitude and direction with respect to it (such systems are called inertial), and that the speed of light is independent of the motion of its source (it is constant regardless of the observer's state of motion or the source). The special theory of relativity is thus based on the assumption of the absolute and invariable nature of the speed of light c and on the assumption that no physical disturbance can travel faster than c.

If the speed of light is indeed constant regardless of the observer's state of motion what does this mean for the nature of time? If time is absolute the speed of light cannot be constant, and if the speed of light is measured (or deduced from Maxwell's equations) to be constant, time cannot be absolute. As Lindley writes,[41]

> "The idea of absolute, universal time, agreed to and used by all came from Newton not because he did experiments to establish how time flowed but because in order to set out a mathematical quantitative theory of motions and forces he made the obvious, simple assumption. There was nothing in Newton's physics that proved the existence of absolute time; it followed from no empirical analysis, nor from any more fundamental assumption. It was an axiom, something taken to be true because it seemed right and led to a theory that worked."

Indeed, special relativity has changed all this and has provided a new set of rules which enable understanding of how time measured by one person relates to the time measured by another and, thus, how the laws of nature relate to the relative motion of the observers who established them. Absolute time is no longer that way. Time is a relative quantity. The rate at which time passes depends on one's motion. A general scientific truth thus emerged: the laws of nature remain the same regardless of the observer's relative motion. They are the same for all observers, identical no matter who describes them, "no matter when and where they are established."[42]

Not only the classical concept of time has been revised, but also those of space, mass, and energy. In the new theory the speed of light in vacuum c appears as a universal constant, its fundamental character transcending its historical connection to electromagnetism. It relates space to time in the Lorentz transformation (Equations (8) and (9)), and appears in the connection between velocity, energy, and momentum of a body having at rest the mass m_0. The concepts of absolute time and absolute space of Newtonian mechanics are now replaced by the postulates of two new absolutes: *the speed of light c is constant and the speed of light c cannot be exceeded.*

Most significantly, special relativity has unified our concepts of mass and energy by showing the *equivalence of mass and energy.* The mass *m* of a body -- defined as the ratio between the force exerted on the body and its acceleration -- becomes variable with the body's velocity. It is a relative quantity dependent on the inertial system to which it is referred. If the body is moving with a speed υ with respect to the reference system, its mass relative to that system is given by

$$m = m_o/[1-(\upsilon/c)^2]^{1/2}. \tag{10}$$

The mass m_0 is known as the body's *rest mass* and can be regarded as a measure of the amount of matter in the body as distinct from its *inertial mass m* which determines its reaction to a force. If $\upsilon \ll c$, the inertial mass is nearly equal to the rest mass m_0, but *m* becomes very large as the particle speed approaches the speed of light (for instance, when $\upsilon/c = 0.999$, $m/m_0 = 22.4$; for massless particles, such as photons, $m_0 = 0$, $E = pc$, $\upsilon = c$) making it impossible within this theory to accelerate a body by application of a finite force to speeds exceeding the speed of light *c*. The energy of the body with respect to the same inertial system is

$$E = m_o c^2/[1-(\upsilon/c)^2]^{1/2} = m c^2. \tag{11}$$

The body's relativistic kinetic energy is the increase in its mass arising from its motion. A body has kinetic energy when its *m* exceeds its m_0.

When $\upsilon \ll c$,

$$E = m_o c^2 + 1/2\, m\upsilon^2. \tag{12}$$

This corresponds to the classical value plus an amount $m_o c^2$ associated with the rest mass m_0. Thus, energy and mass are to be thought of as equivalent, as different aspects of the same quantity. An energy *E* is equivalent to a mass E/c^2 and a mass *m* is equivalent to an energy mc^2. In classical mechanics although the total amount of energy in a closed system (a system which is not in interaction with its surroundings) is constant, transformations may occur between different kinds of energy. These transformations should now involve the energy of the rest mass. For instance, the transformation of the energy of the rest mass into kinetic, or radiant, or other form of energy would mean the annihilation of matter. In the same way the inverse transformation -- the creation of matter -- would be possible. This seemed unbelievable at the time the theory of special relativity was advanced. Today, it is very common in modern physics. For instance, we know that the enormous amounts of energy which are released from uncontrolled nuclear explosions and from controlled nuclear fission in power reactors come from the annihilation of matter. We also know this for many physical processes such as electron-positron annihilation. Energy and mass are equivalent quantities, two forms of the same entity the nature of which we still do not know. Modern science is still trying to find out the nature of mass through studies of elementary particle reactions (Chapter 4).

Furthermore, the classical laws of conservation of mass, energy, and momentum are now revised. While in classical physics the mass, energy, and momentum of a closed system are separately conserved (that is, the total amount of each within the system remains constant), in the new theory these three conservation laws can be retained only if mass, energy, and momentum are redefined. Because mass and energy are equivalent and interchangeable there are now no longer separate conservation laws of energy and mass. Instead, these are combined -- united -- into a single and simpler conservation law, the law of conservation of mass-energy which is invariant under a Lorentz transformation. This was necessary for the extension of man's knowledge from the macrocosmos to the microcosmos and from the slow to the fast objects. Indeed the subsequent demonstrated validity of these generalized conservation laws (e.g., from the balance of energy in an atomic bomb or a nuclear reactor), along with the demonstrated variation of m with υ (e.g., the dynamics of energetic particles, such as relativistic electrons, in accelerators), and the direct observations of time dilation (e.g., in μ-meson decay) constitute experimental confirmation of the theory and its practical consequences. Nonetheless, like any other physical theory, special relativity is a set of rules based on a number of crucial assumptions. Its postulates are assumed valid as long as predictions based on them conform to the facts. They will change when exceptions are found (if, for instance, it can be shown that the speed of light can be exceeded or is direction dependent) and they will have to be replaced with yet other postulates, in the perennial evolution of man's scientific comprehension of physical reality.

Indeed ten years after special relativity, in 1915, Einstein formulated his general theory of relativity which is capable of coping with forces and accelerations, systems possessing strong gravitational fields, and objects moving very close to the speed of light. The new theory revised further man's concepts of space and time and Newtonian ideas of gravity. Space is not flat but curved, the new theory said. This curvature of space is generated by the presence of mass so that a heavy object causes a depression into which other bodies are drawn. There is no longer a force of gravity in the old Newtonian way to act instantaneously. Instead, mass distorts space and bodies traveling through curved space do not travel in perfectly straight lines but along curving trajectories. Mass curves space only locally and space assumes structural and physical characteristics. The connection between gravity and the curvature of space offers a new understanding of what mass *means* (not, however, of what mass *is*). Mass causes curvature and curvature influences the paths of masses. The concepts of inertial and gravitational mass are united. The action at a distance is replaced by a gravitational influence that spreads though space at the same speed as electromagnetic radiation. And just as Maxwell's equations yielded solutions that describe electromagnetic waves traveling in space, "in general relativity there appeared to be gravitational waves, in the form of ripples in space itself, traveling at the speed of light."[43] However, while the existence of electromagnetic waves has long been demonstrated there is as yet no direct proof of the existence of gravitational waves [44] (see Chapter 4).

3.6. THE MODERN ERA

3.6.1. The Microcosmic Universe

By the latter part of the 19th century the atomic constitution of matter was well established. The ninety elements then known were considered to be the indivisible elemental building blocks of all matter and nothing beyond. Just as this notion was being firmed up and began aiding the understanding of diverse phenomena in chemistry, physics, and other branches of science such as medicine and biology a new frontier opened up: that of subatomic particles. The atom is not the ultimate indivisible unit of matter: the cosmic onion[45] has inner layers.

The subatomic era began with the discovery of X-rays in 1895 by Wilhelm Konrad Roentgen. Its opening stages commenced with rapidity:

the discovery of radioactivity in 1896, the discovery of the electron in 1897, the discovery of radium in 1898, the discovery of energy quantization in 1900, the discovery of γ-rays in 1900, and so on. From this time onward, layer after layer the cosmic onion's structure was unraveled and with it the constitution of matter and the intimate connection between matter and radiation. This was basically accomplished by following two distinct paths: the path of light and the path of subatomic particles. We shall refer only to those aspects of light and particles which help us complete our historic perspective.

3.6.1.1. Electrons

The discovery of the electron[46] and its recognition as a common constituent of all atoms and as one of the most fundamental, abundant, and reactive particles in nature, was gradual. While the Periodic Table of the Elements pointed to common structural units in the atoms of the different elements, the discovery of the electron had its origins in the study of electrolysis. The consequences of the passage of electric current through liquids and Faraday's laws of electrolysis (1833) provided the first evidence about the smallest quantity of electric charge carried by electric charge carriers and the ratio e/M of the fundamental charge e to the mass M of the charge carrier.[47] However, it was the consequences of the passage of electric current through gases and the associated phenomena of gas discharges,[48] that ultimately led to the discovery of the first subatomic particle, the electron, by J. J. Thomson in 1897 just as such studies led to the discovery of X-rays by Roentgen a year earlier. Thomson was studying the negatively-charged particles moving towards the anode in an electric gas discharge. He was able to measure the charge-to-mass ratio, e/m, of these anode-directed charged particles, which he found to be over 1,000 times greater than that for the lightest carriers of charge in electrolysis. Thus, the mass of these negatively charged particles was less than one thousandth the mass of the hydrogen atom. Subsequently, the electronic mass and electronic charge were measured separately, first roughly by J. J. Thomson and then precisely by R. Millikan.[49]

3.6.1.2. Prelude to the Discovery of the Nucleus

The road to the discovery of the nucleus opened up when it was realized that the atom must contain positively charged constituents. Since atoms exist and are neutral, and since electrons are constituents of all atoms and are negatively charged, there must also be positively charged constituents in atoms and in such number as to balance with their charge the negative charge of the atomic electrons. Evidence for the existence of positively charged atomic constituents came from a number of sources. One such source was radioactivity, which was discovered accidentally by Becquerel in 1896 in his effort to understand the nature of X-rays discovered (also accidentally) by Roentgen the year before. Becquerel observed that unexposed potassium-uranium-sulphate emitted radiation which could penetrate black paper and fog a photographic film. The understanding of this radiation had to await about half-century, but the signal was picked up in 1896: *the nuclei* of the uranium atoms in the potassium-uranium-sulphate were spontaneously breaking up releasing the observed radiations. Indeed, experiments by Ernest Rutherford (1871-1937) between 1896 and 1900 showed that Becquerel's radiation contained three distinct components: alpha (α), beta (β), and gamma (γ).

Gamma radiation was electromagnetic radiation of much higher energy than X-rays. Beta radiation consisted of energetic negatively charged particles, electrons. Alpha rays were massive positively charged particles. Rutherford's systematic study of the nature of the rays emitted from radioactive substances showed that helium gas was generated by radioactive substances which emitted alpha rays. This was a profound experimental discovery which led to an equally profound conclusion: alpha rays are charged helium atoms. Inductively then Rutherford uncovered a profound scientific truth: atoms break up, disintegrate spontaneously,[50] and radioactivity is a demonstration of the spontaneous break up of one atom into another. This is a classic example of the inductive method of science. The researcher was guided by no theory or pre-existing construct. *There was only the scientist observing nature passionately, fighting with himself and the limitations of his interrogating technique, contemplating the meaning of his observations, and carefully but boldly edging toward the truth. A beautiful intimate dance with mystery! As elegant, in the author's view, as Einstein's axiomatic deduction of special relativity.*

The next natural step was the search for an answer to the question: How are these alpha particles formed? In search for an answer to this question the scientist now had a new tool: the alpha particles themselves. A new discovery uncovered a new scientific truth, raised new questions, and provided new tools for the next level of discovery, the beginning of man's first look inside the atom.

3.6.1.3. The Discovery of the Atomic Nucleus

Once radiation from radioactivity became available what did the scientist do? He used it as an instrument, as a new tool to study matter in much the same way he used earlier the electric current. Especially useful for probing inside the atom and learning about atomic structure were the alpha particles. After all, they came from inside the atom. At this time (ca 1910) there was general agreement that atoms contained a small number of electrons so that most of the atomic mass resides in the positive charge. But how was the positive charge distributed in the atom? Was the atom like a sphere of positive charge in which the electrons were embedded, like J. J. Thomson thought, or was the positive charge contained within a central massive nucleus as E. Rutherford envisioned? Two different models indeed. In Rutherford's view the electrons would move in orbits around the nucleus under its electrical attraction just as the planets orbit the sun under the gravitational attraction. In Rutherford's model therefore *condensed matter would be mainly empty* since virtually all atomic mass is concentrated in the tiny nucleus. Such differences would manifest themselves in the behavior of alpha particles impinging on a metal foil.

According to the Thomson model, when the alpha particles impinge on a thin metal foil they would all pass through it and be deflected at small angles from their original direction. According to the Rutherford model, when the alpha particles impinge on the metal foil their preponderance would pass through undeflected, but a very small fraction -- those α-particles that would pass close to an atomic nucleus -- would suffer a large deflection. In 1909 Ernest Marsden, a student of Rutherford's, reported that when he bombarded a thin wafer of gold with alpha particles most of the alpha particles indeed passed straight through but about one in 10,000 bounced back. "Can cannon-balls recoil from peas?" Rutherford exclaimed in amazement. It was the most incredible event that had happened in his life he remarked. "It was as if you had fired a 15-inch shell at a piece of tissue paper and it came back and hit you."[51] But the conclusion was unavoidable: *somewhere in the gold atoms must be concentrations of matter much more massive than alpha particles.* An amazing finding was picked up by the skillful eye of a student. The observation must now be repeated, scrutinized, quantified, interpreted. Indeed, in 1911 Rutherford announced his solution to the puzzle and his construct: *the model of the atomic nucleus.* He proposed that all of the atom's positive charge and most of its mass are contained in a compact nucleus at the center of the atom. The nucleus occupies only a very small fraction (about 10^{-12}) of the atomic volume -- hence the very small number of violent collisions with α-particles -- and the electrons are spread around outside. More experiments on the fraction of alpha particles incident on a variety of substances which were deflected through a given angle convincingly showed that the nuclear model was correct. This was a landmark in man's understanding of the structure of the atom.

The discovery of the atomic nucleus is another example of the step-by-step advancement of science: the discovery of alpha particles, their identification as helium positive ions, the realization of their value as a tool to study the atom, the observation of an anomaly in their scattering pattern, the inductive inference of the atomic nucleus, and the building up of a new construct: *the existence of the atomic nucleus and the prodigious emptiness of matter at the atomic level.* And now the next level: *the structure of the atom.* How is this atomic system held together and how can a structural construct be fathomed to account for the observed emission of light, X-rays, alpha, beta, and gamma radiation, and so on from the atom?

3.6.1.4. The Atomic Electrons and the Atomic Nucleus

How are they kept together? If, as Rutherford suggested, the electrons orbited around the central nucleus classically the atom could not be stable: the electrons would radiate energy and spiral down into the nucleus. Once again the scientist was faced with a new reality and a recurring problem: the existing theory could not account for the facts, in this instance for the stability of the atom. The laws of the macrocosmos seemed inadequate to explain the atom of the microcosmos. The scientist was "cornered" yet again in his description of nature: his understanding was incomplete and contradictory. Time for a new construct to explain the new facts about nature. And this began with Niels Bohr in 1913.

As scientists before him -- and as scientists after him -- Bohr surveyed the existing facts and ideas of description of the new properties of matter and used them as a basis for new propositions. Thus, Planck's discovery that radiant energy is quantized led Bohr to propose that the energies of the electrons in atoms are also quantized: *electrons in atoms can have only certain pre-described energies* controlled by the same fundamental entity, Planck's constant h. Bohr postulated further, that in the normal state (the lowest-energy state or the ground state in modern terminology) the electron is circulating in the orbit for which the principal quantum number $n = 1$ and that in that state the atom is stable, that is, it does not radiate energy. Electrons circulating in allowed higher-energy orbits (orbits with $n > 1$) were similarly postulated not to radiate energy in spite of their higher energies. However, electrons in higher-energy, E_i, states could jump to a lower-energy, E_f, state and in such a *transition* the excess energy was assumed to be emitted as electromagnetic radiation of frequency v given by the Planck-Einstein relation $hv = E_i - E_f$. Bohr applied his idea to the simplest atom, that of hydrogen (H) whose nucleus is orbited by one electron, and predicted the wavelengths of the radiation emitted from excited H atoms in agreement with the measurements. Thus, Planck's quantum theory that was applied successfully to radiation by Einstein had now been successfully applied to matter by Bohr.

The ways the atomic spectra are modified when atoms are located in magnetic or electric fields were explained by hypothesizing that not just the energy, but also the angular momentum of the orbiting electrons is quantized. The concept of quantization was thus extended to the physical quantity of angular momentum. The magnitude of the angular momentum for an electron in an orbit around the nucleus must be restricted to integer multiples of Planck's constant h. A little later, the electron was discovered to have an intrinsic spin of its own of magnitude ½ h, and gradually a series of quantum numbers (the principal quantum number n, the angular momentum quantum number l, the spin quantum number s, and the magnetic quantum number, m) were advanced for a fuller prescription of the electron orbits and the absorption and emission of radiation by atoms. Ultimately, Bohr's model of the atom was replaced by a more complete description that came later with the full development of quantum mechanics.

3.6.1.5. *The Wave Nature of Matter*

Like the conservation laws, symmetry guided scientific discovery in a supreme way. A classic such example is Louis de Broglie's extension in 1924 of the idea of the wave-particle duality. If, he argued, light sometimes behaves like a wave and sometimes like a particle, those entities thought to be particles (e.g., electrons, atoms, other objects) might also have wave properties. He, then, ascribed a wavelength λ to a particle of mass m and speed v as

$$\lambda = h/mv. \tag{13}$$

Large slow-moving particles have very short wavelengths compared to their dimensions, while small fast-moving particles have long wavelengths. For example, a ball weighing 200 g and moving with a speed of 20 m s^{-1} has a de Broglie wavelength of about 1.7×10^{-34} m, while an electron, whose rest mass is 9.108×10^{-28} g, moving with a speed of 5.9×10^6 m s^{-1} (this corresponds to a kinetic energy of 100 eV) has a very much longer de Broglie wavelength of 1.2×10^{-10} m = 1.2 Å. The wavelengths of electrons were first measured in 1927 by C. J. Davisson and L. H. Germer. Electrons were shown to be diffracted just like photons (X-rays) of the same wavelength when they were scattered from a metal crystal since the spacing of atoms in the crystal is on the order of 10^{-10} m and the orderly array of the atoms in it serve as a type of diffraction grating. Electrons impinging on a metal exhibited a diffraction pattern consistent with a diffracted electron wave having a wavelength just as predicted by the de Broglie Equation (13). Later, experiments showed that other particles such as protons, neutrons, and atomic ions show wave-like properties, but their wavelengths are shorter because they are more massive. In fact, the wave-particle duality applies to all material objects as well as to light. It is true of all matter. However, because of the smallness of the quantum of action h (Planck's constant) and the bigness of a large body's mass, the wavelength of a large body is completely insignificant compared to its dimensions. Practically, then, the wave-particle duality is an element unique to the microcosmic universe.

The wave-particle view (Table 3.7) of physical objects offers complementary descriptions of one and the same reality. By being aware of both the particle and wave aspects of matter, man's understanding of matter is more complete than it would be by either view alone. And yet one could still justifiably ask: so, what is an electron? No one has actually seen an electron *directly* since it was first discovered by J. J. Thomson. And yet experiments since that time were repeatedly interpreted as being caused by tiny negatively charged particles we call electrons. Experiments such as those involved in the passage of electricity through gases, electron flow in a wire, electron diffusion and drift in gases or inside a metal, electron collisions in gases and discharges, photoelectron emission from materials, electron emission from a hot filament, electron attachment to molecules, electron detachment from negative ions, and electron transfer from one atomic species to another are all valid pictures and they are all consistent with the particle description of the physical entity we call electron. Yet, electrons possess properties that are interpretable only in terms of waves.

As we have mentioned earlier in this section, when electrons strike the face of a crystal they are diffracted by the regularly spaced atoms of the crystal just as are light waves. Interestingly, also, when free slow-moving electrons collide with other particles -- atoms or molecules -- it is often found that they interact with them far more strongly than if these interactions were limited to their coming in contact as particles (geometric cross sections). For instance, slow electrons are captured by molecules

Table 3.7. Particle-wave duality[a] and complementarity[b]

Particle/Wave relations	Explanation
$E = h\nu$	Light particles (photons; quanta)
$\lambda = h/m\upsilon$	Particle waves
$\Delta x\, \Delta p \geq h/2\pi$	Uncertainty principle[c]
$H\Psi_n = E_n\Psi_n$	Wave (Schroedinger) equation[d]
$|\Psi_n|^2$	Statistical causality[e]

[a] Duality is true of all matter.
[b] Important in "all" life.
[c] A consequence of duality rather than a limit to accuracy.
[d] The equation is solved for a particular potential and searched for values of E_n for an assumed Ψ_n.
[e] See text.

forming negative ions with probabilities (cross sections) that are determined by the size of the wavelength of the electron waves. The author's work[52] on electron capture by molecules showed that electrons with kinetic energies less than about 1 eV are attached to molecules with cross sections determined by their electron de Broglie wavelengths just as Equation (13) prescribes, rather than by considering the electrons as particles. Interestingly, Equation (13) predicts, and experiment shows, that slow electrons are more reactive than fast electrons in their interactions with matter. Slow electrons "see" better than fast electrons do. Similar statements can be made about neutrons which are slowed down by moderators to increase their reactivity.

It is thus evident that experiments are some times best interpreted by using the particle model and sometimes by using the wave model. This extra freedom the scientist gave to himself represents a strength and a weakness in his conceptualization of natural phenomena. A strength in that he has achieved a new level of construct unification, that of the *classical* concepts of particle and wave. A weakness in that the two pictures, particle and wave, are merely two ways employed to extrapolate from the macroscopic to the microscopic world. Models which do not necessarily reflect reality but which at present are useful in describing the observed behavior of a physical entity. The electron is neither a wave nor a particle but rather an entity defined by its measured properties which can best be described by viewing it sometimes as a wave and sometimes as a particle whichever way best approximates what is observed. Describing electrons, and microcosmic matter in general, this way -- sometimes as particles and sometimes as waves -- may seem paradoxical, but our present description of nature's behavior is better off and more complete with it.

In ways similar to the past, in the mid 1920's the time had come for a synthesis of all these new ideas into a new paradigm: *wave mechanics.* This time the new formalism originated in 1926 with Erwin Schroedinger and has developed since into the elegant structure of quantum mechanics by innumerable others. T*he motion of an electron in an atomic system could be understood by investigating the wave motions associated with it.* For a particle -- free or under the influence of a known force -- a wave equation could be set up such that its solutions would give the permitted energies (eigenvalues E_n) and the corresponding wavefunctions Ψ_n. Particles no longer have separate velocities and well-defined positions, but instead they have *a quantum state* which is a combination of their position and velocity. From a knowledge of the wavefunctions Ψ_n describing the quantum states of the particle, the probability of finding it in a confined region of space at a given time could then be determined. It is proportional to the square of the amplitude, $|\Psi_n|^2$ of the associated wavefunction Ψ_n in that region. Thus, when electrons are confined in the finite region of space around the nucleus they form specific shapes and patterns characteristic of waves. Some of these are shown[53] in Figure 3.1 for a number of values of the principal quantum number n, the angular momentum quantum number ℓ, and the magnetic quantum number m. In the figure are shown probability density distributions $|\Psi_n|^2$ for electrons in the lowest stationary quantum states of the H atom. The probability of finding the electron at any point is proportional to the brightness of the picture. The proton is at the center of the pattern in all cases and the magnification is $\sim 10^{-8}$.

These pictures of "probability density" do not change with time. The electron does not radiate when it occupies these stationary quantum states.

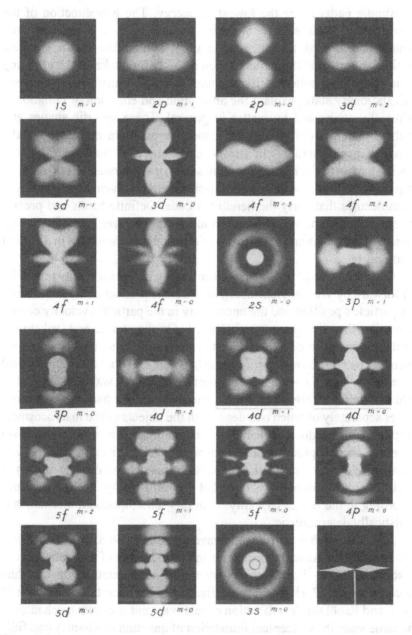

Figure 3.1. Patterns of electron wave functions in various quantum states of the hydrogen atom (from Ref. 53). Transitions between such stationary states explain the absorption and emission of light by atoms, and illuminate our knowledge of the energy exchanges between radiation and matter.

The simple patterns are the lowest in energy. The wavefunction of the ground state, with energy quantum number $n = 1$ is smooth and symmetrical. Indeed, the electrons in atoms assume the lowest, simplest possible patterns and this explains their stability for it takes energy to jump to the next higher pattern. Discontinuous jumps - - transitions - - between such stationary states explain the absorption and emission of radiation -- light -- by atoms and by matter in general. These specific shapes and electron patterns (defined by the respective quantum numbers) are the primal shapes of nature.[54] In a truly remarkable way they determine the specific properties (chemical and physical) of the atoms. All of nature's shapes can be fundamentally traced to such patterns, even the properties of the molecules that carry the hereditary code. Definite forms and precise dimensions are imposed on the multitude of atomic aggregates of nature by the size of the atoms and the patterns of electron wavefunctions in different quantum states.

A consequence of the particle-wave duality is the uncertainty principle, put forward by Werner Heizenberg in 1927: the product of the uncertainty in a particle's position and the uncertainty in the particle's velocity cannot be smaller than the quantum of action h. On the atomic and subatomic level the exact position and the exact velocity (or momentum) of a particle can never be *both* known *simultaneously*. There exists an inherent physical uncertainty when a switch is made from a particle to a wave description of nature which is clearly absent when one speaks about particles and about waves separately or when one deals with the objects of the macrocosmos. In spite of this irreducible element of uncertainty at the microcosmic level of reality, there is precision at the quantum level. Born expressed this and the essence of wave mechanics in the beautiful statement: "The motion of particles follow probability laws but the probability itself propagates according to the law of causality."[55] In this sense, quantum mechanics is statistically deterministic.

And so through Schroedinger's wave equation in 1926, Born's statistical interpretation of quantum mechanics and introduction of the probability of a state in 1926, Bohr's complementarity interpretation of quantum mechanics in 1927, Heizenberg's derivation of the uncertainty principle in 1927, and Paul Dirac's unification of relativity and quantum mechanics in the same year, the conceptual foundation of quantum mechanics was fully in place. Quantum theory since expanded into numerous directions and reached unimaginable levels of sophistication and complexity. In its evolution the computer has been an indispensable partner.

3.6.2. Pealing the Inner Layers of the Cosmic Onion

3.6.2.1. Protons and Neutrons

By the early 1930's the electronic structure of the atom was broadly understood: how the atomic electrons are distributed among the atomic energy levels and the relation of this electron distribution in the atom to the chemical properties of the elements and their regularities (as were first noticed by Mendeléev 60 years earlier). Chemistry was being understood through physics. As to the atomic nucleus, the then state of knowledge was simple: the nucleus was just the minute, very dense and positively charged entity in the atom's center. This was soon to be changed in large part due to the work of Rutherford on the scattering of alpha particles from various nuclei. During the first world war, Rutherford used alpha particles from radium to bombard nitrogen gas and in 1919 he published his results establishing beyond doubt that fast protons were produced in collisions between alpha particles and nitrogen nuclei. Helium and nitrogen, he showed, were transformed into hydrogen and an isotope of oxygen, viz.,

$$^{7}N_{14} + {}^{2}He_{4} \rightarrow {}^{1}H_{1} + {}^{8}O_{17}. \tag{14}$$

This is the first artificial nuclear reaction obtained using particles from natural radioactivity. The alchemist's dream finally came true. Aristotle's unchangeable atoms are changeable. Modern nuclear physics is born.

Experiments with other types of nuclear targets showed that nuclei could be transmuted into one another and that the nucleus itself has internal structure. Moreover, since in these experiments protons had been ejected from the nucleus of nitrogen and the nuclei of other elements, Rutherford suggested that protons might be the particles which carry the positive charge of all atomic nuclei: the hydrogen nucleus carries one proton, the helium nucleus two, the oxygen nucleus eight, and so on. Important as this step had been in terms of nuclear charge, it presented a problem with regard to the atomic mass. For instance, the eight protons of the oxygen atom would indicate that its mass is eight times that of the hydrogen. However, on the basis of the determination of the mass of the chemical elements the mass of the oxygen atom was 16 times that of the hydrogen atom. Could the rest of the nuclear mass be made up by a combination of protons and electrons? Such a proposition was ruled out by other facts such as the nuclear spin.

Once again the existing paradigm was incapable of explaining the experimental facts. In this instance the inductive approach offered an intuitive guess, just as the deductive approach in similar instances had offered a new postulate. Rutherford - - the man who discovered nuclear transformations, the man who discovered the atomic nucleus, the man who discovered the proton as a nuclear constituent - - made a bold suggestion: might there not be another particle in the nucleus as heavy as the proton but without electric charge, a "neutron"? Indeed, in 1932 the other nuclear constituent - - the neutron that Rutherford's intuition guessed - - was discovered by J. Chadwick. The discovery was itself a beautiful example of how a scientist skillfully devices experimental methods for nature to reveal itself. The neutron having no charge is difficult to detect experimentally for it is invisible to most instruments (detection methods in physics normally rely directly or indirectly on light and/or charge for detection). Actually, Irène Curie and Frédéric Joliot had evidence for the neutron prior to Chadwick's discovery but misinterpreted it.[56] They bombarded beryllium targets by alpha particles and discovered that from such collisions came electrically neutral radiation which they mistook as X-rays. Chadwick did the same experiment as the Joliots, and since he could not detect neutrons directly, he conceived of a way to make them reveal themselves indirectly by passing them through hydrogen gas (or, as the Joliots did, through a hydrogen-rich material such as paraffin) and looking at the products of their collisions with the hydrogen atoms. If neutrons were produced when alpha particles collided with beryllium, they would collide with the hydrogen atoms and because neutrons and hydrogen atoms have almost the same mass the likelihood of neutrons knocking protons (H^+) off the hydrogen molecules is large. Protons being charged could then be easily detected. The hydrogen gas (or the paraffin) was the trap that would make the invisible neutrons visible to Chadwick's detectors tuned to look for protons ejected from the gas or the paraffin. On the other hand, if the neutral radiation were X-rays as the Joliots surmised, it would remove electrons from the atoms in the paraffin, but it would not eject protons. Chadwick detected copious quantities of neutrons.

A new subatomic particle was discovered similar in mass to the proton but with no electrical charge. The neutron Rutherford predicted was generated via the reaction

$$^4Be_9 + {}^2He_4 \rightarrow {}^0n_1 + {}^6C_{12}.$$ (15)

With the discovery of the neutron, it was established that nuclei are composed of two types of particles (the nucleons): the neutron ($^{0}n_1$ or n) and the proton ($^{+}p_1$ or p). Based on this step-by-step process of experimental science a new picture of the atomic nucleus emerged: the atomic nucleus is made up of Z protons and A-Z neutrons, Z being the atomic number and A being the atomic mass. A exceeds Z except for the hydrogen atom for which $A = Z$. Since an atom is electrically neutral, the positive nuclear charge will be $+Ze$, equal to the charge $-Ze$ of the atomic electrons. It was soon found that while every nucleus of a given element contains the same number (Z) of protons, it may have different numbers (N) of neutrons and thus every element can have a number of *isotopes*.[57] Isotopes became crucial for the furtherance of man's understanding of nature and for numerous applications stretching from chemistry, to biology and medicine, metallurgy, materials, and industrial processing.

It is perhaps instructive at this point to refer to the decay of the neutron because it is a prototypical example in its demonstration of the significance of the conservation laws in discovery, especially in the discovery of many of the elementary particles of nature. In his effort to understand the phenomenon of neutron decay the scientist relied (as he did innumerable other times) on the validity of established law, the law of conservation of energy, and discovered a new particle, the neutrino. The neutron decay ($^{0}n_1 \rightarrow {}^{+}p_1 + e$) is known as beta-radioactivity and is the source of many transmutations. This decay seemed to be in conflict with the law of conservation of energy (the generalized form which includes the conservation of mass and energy): the energy of the proton and the electron seemed to be less than they should be. Taking as a given that the law of conservation of energy must be valid, Wolfgang Pauli postulated in 1931 that along with the proton and the electron, a new particle, the neutrino (v), is generated in the decay of the neutron in order to explain the otherwise anomalous energy loss. Pauli's predicted particle was observed directly 1956. The neutrino carries no electrical charge and interacts weakly with matter and its detection is difficult. Its existence was inferred by observing the products of its interaction with matter (cadmium) in much the same way the neutron had revealed itself by ejecting protons from a hydrogen gas or paraffin wax.

The discovery of the nucleus and the unraveling of its structure that first began with the finding that nuclei spontaneously decompose and was subsequently followed by the finding that nuclei artificially decompose by bombardment with alpha particles from these natural decompositions, was posed for a new tempo in the decade of the 1930's. Rather than relying on nature to provide the particles to use in his studies of the nucleus, man now began constructing machines causing atomic particles (positive ions) to be accelerated to such high energies that he can use them to decompose nuclei *artificially*. Man made his own atomic bullets. Nuclei were first split artificially in 1932 by John Cockcroft and Ernest Walton who used electric fields to accelerate protons to high speeds, and then directed them at lithium nuclei producing two He nuclei ($^3Li_7 + {}^1H_1 \rightarrow {}^2He_4 + {}^2He_4$). Protons being lighter ions than alpha particles can be accelerated to higher speeds. Cockcroft and Walton made the first practical nuclear particle accelerator and opened up the new era of experimental probing of the nucleus, its internal structure, and, subsequently, the structure of the nucleons themselves.

3.6.2.2. Nuclear Fission

Now that the scientist discovered a new particle, the neutron, the next step was obvious. Use it to further the study matter. Another particle, another series of investigations using the particle itself as a probe. Since, moreover, protons and neutrons are the common constituents of all nuclei, *one nucleus can transmute into another by absorbing or emitting these particles.*

The bombardment of naturally occurring nuclei with neutrons to form new isotopes by neutron capture by the target nuclei was an exciting field in the early 1930's. Since the probability of neutron capture by the nucleus in such collisions is larger the slower the neutrons, Enrico Fermi first slowed down the neutrons by passing them through paraffin and then allowed them to collide with the nuclei under study. In this manner he was able to attach neutrons to a total of forty two different nuclear targets producing new artificial isotopes, that is, he *artificially modified the nuclei of various atoms.* By 1934, Fermi came to the then heaviest known element -- the uranium atom -- and there one of the greatest physicist that have ever lived, "made a mistake for which we may be thankful."[58] When Fermi irradiated uranium he observed "some puzzling phenomena" which he presumed to be evidence for the production of a new isotope of the uranium atom or the first transuranic element[59] (a new element one place above uranium in the Periodic Table of the Elements). And so, in 1934, so dangerously close to WWII, Fermi missed the real explanation of his experiment: nuclear fission. Man was lucky!

The work of Enrico Fermi, Otto Hahn, and Lise Meitner in the 1920's and 1930's led to the discovery of the nuclear fission reaction by Hahn and Fritz Strassmann in 1938. When they bombarded uranium with neutrons instead of a new element they found the formation of barium, an element with a nucleus of 56 protons. Slow neutrons split, fissioned, the uranium nucleus into two almost equal halves via the reaction

$$^{92}U_{235} + {}^0n_1 \rightarrow {}^{92}U^*_{236} \rightarrow {}^{56}Ba_{144} + {}^{36}Kr_{89} + 3\,{}^0n_1. \qquad (16)$$

In Reaction (16) the asterisk indicates that the $^{92}U^*_{236}$ "compound nucleus" formed by neutron capture by the $^{92}U_{235}$ nucleus has excess energy. In Reaction (16), also, the number of neutrons and protons is separately conserved. Within days of the announced discovery of nuclear fission the result was confirmed by other groups.

In a nuclear fission reaction not only the heavy compound nucleus fissions into two parts of comparable mass, but additional slow neutrons are produced which can trigger the break-up of other uranium nuclei. Enormous amounts of energy can be released in the ensuing "chain reaction." The essential ingredients of the atom bomb were thus in place in Italy and Germany up to five years before WWII. And the world didn't know it. This is perhaps an instance when man's ignorance was best for him.

3.6.2.3. The Steps Toward the Atomic Bomb[60]

The fission fragments in Reaction (16) necessarily have excess neutrons compared to the stable nuclei of the same atomic number. They can eliminate this imbalance either by the slow process of beta decay or by direct neutron emission. In the second case the secondary neutrons may be used to produce new fissions, and if they are in sufficient number, they produce more neutrons than in the first generation. In this way a *divergent* nuclear chain reaction is achieved. If the chain reaction occurs very rapidly in an uncontrolled way, a violent explosion follows and one has an atomic bomb or strictly speaking a nuclear bomb. On the other hand, if the reaction can be controlled and can be brought to a stationary state, a new power source, the nuclear reactor, is achieved. Both paths lay open: the path of the atomic bomb and the path of nuclear power. The destructive and constructive aspects of scientific knowledge archetypical in their power for either good or evil.

The step from the discovery of nuclear fission (1938) to a sustain nuclear chain-reaction (1942) took only four years. The first critical nuclear reactor was started at Stagg Field in Chicago on December 2, 1942. The race from a discovery in physics to the construction of the first atomic bomb essentially began with Einstein's letter to president Roosevelt in August 1939 (Chapter 2) and the Manhattan District Project was organized in 1942 with the sole purpose to build the atom bomb once the political decision to do so had been made. While this was taking place in the United States similar efforts were started in the United Kingdom, France, and Germany. Early on in this effort it was realized that a uranium bomb required separation of the isotopes $^{92}U_{238}$ and $^{92}U_{235}$. Isotopic separation of uranium was thus one way to make an atomic bomb (the uranium bomb). Another way was the formation of sufficient quantity of plutonium to make a plutonium bomb. The former required mass spectrographs or gaseous diffusion to separate the fissile material, and the latter a nuclear reactor to produce it and chemical processes to separate and purify it. Both atomic explosives, $^{92}U_{235}$ and $^{94}Pu_{239}$, were produced and science and technology worked out the details of building the actual device. It might be noted that just as in the case of uranium four years lapsed between the discovery of plutonium and the first plutonium-based atomic bomb. The first three atomic bombs were built at Los Alamos, New Mexico. The first atomic bomb was detonated (Figure 2.1) at dawn July 16, 1945. The other two bombs were dropped on Japan (Table 3.8). The second world war ended soon thereafter, but man's struggle with his consciousness and with his predicament continued unabated. *Neither the world nor science has been the same since.*

Table 3.8. Chronology of the atomic and hydrogen bombs[a]

Event	Date
Discovery of nuclear fission	1938
First sustained fission reaction	1942
First nuclear explosion in New Mexico, USA	July 16, 1945
The uranium bomb ("Little Boy") detonated over Hiroshima, Japan	August 5, 1945
The plutonium bomb ("Fat Man") dropped on Nagasaki, Japan	August 9, 1945
First explosion of the hydrogen bomb in the Marshal Islands	October 31, 1952

[a] The chronology refers to events which first occurred in the United States. Similar events followed in other countries.

3.6.2.4. Nuclear Fusion and the H-bomb

Just as the scientist had found that the very heavy nuclei fission when they are bombarded with slow neutrons and release enormous amounts of energy in the process, he also found that the very light nuclei fuse into bigger nuclei when they are brought very close to each other and release even larger amounts of energy when fused. We elaborated on some of these reactions in the previous chapter when we mentioned the basic reactions that can be used to produce fusion energy. While both controlled and uncontrolled nuclear fission energy had been accomplished, to date only uncontrolled fusion energy has been achieved and employed to make yet a more powerful bomb: the H-bomb. Table 3.8 lists some dates in the chronology of these developments.

Cosmic fires in the hands of man. All too concentrated form of energy, threatening man and nature in its uncontrolled release.

3.6.2.5. The Innermost Layer of the Cosmic Onion

Following the discovery of the atomic nucleus using nuclear particles from natural radioactive decay, nuclear reactions were studied using particles accelerated by machines (linear and cyclotron accelerators) constructed by man. By the 1960's huge accelerators were built, some several miles long, which are capable of accelerating electrons or protons to speeds approaching the speed of light. For instance today the Stanford Linear Accelerator in California accelerates electrons to energies over 20 GeV and the Fermi National Laboratory Collider in Batavia, Illinois, accelerates protons and antiprotons to a total energy of 1.8 TeV (see Chapter 4). As the energy of the accelerated particles was increased, these big machines shifted the study of the nucleus toward nature's most elemental reactions. The collisions of such energetic particles with nuclear targets allowed them to penetrate into the neutrons and the protons of the nucleus enabling a study of their own inner structure. Neutrons and protons were found to consist of more basic particles called quarks. A powerful force clusters quarks together building the neutrons and protons of the nuclei. As acceleration machines allowed more and more energetic projectiles (mostly electrons and protons) more and more "elementary" particles were discovered and smaller and smaller sizes of matter were probed.

From natural radioactivity's alpha, beta, and gamma radiation, to electrons, photons, protons, and neutrons, to W and Z bosons and quarks, the scientist discovered an ever increasing array of elementary particles - - some found subsequently to be not so elementary - - each intimately connected to the deepening of his understanding of the physical world. In a nearly one-hundred-year long step-by-step effort the scientist sought to uncover the ultimate structure of matter, identify the fundamental particles from which the cosmos is built of, and formulate the laws which govern the combinations of the fundamental particles and bind them together into nuclei, atoms, molecules, the objects we see, the earth and the stars, the galaxies, the cosmos. In this nearly one-hundred-year-long frontier the scientist attempted in an almost paradoxical way to explain the very big - - how matter and the cosmos were formed - - by developing an understanding of the very small, the elementary particles of nature. The scientist had realized that to understand the forces of nature he has to understand the nature of the elementary particles because these are the entities which are exchanged between interacting matter, the agents which "mediate" the forces of nature. By understanding the reactions of these elementary particles man ultimately hopes to unify his view of the forces of nature (Chapter 4). That is a level of unification that is still to come.

In Figure 3.2 is schematically illustrated the penetration of science into smaller and smaller units of matter.

3.6.2.6. Scientific Milestones During the First Half of the Twentieth Century

In closing this section we list in Table 3.9 examples of the outstanding scientific accomplishments during the first half of the 20th century.

3.7. THE CHANGING CHARACTER OF SCIENCE

The scientific revolution took a marked turn the first half of the 20th century that resulted in the unavoidable marriage of science and technology. During this period, but especially after WWII, science had become an indispensable element for the many profitable phases of technology (Chapter 7).The practical benefits of scientific discovery accelerated interest and investment in science. While in Europe government support of research on a broad basis began in the 1920's, by and because of the developments during WWII, the USA and the Former Soviet Union

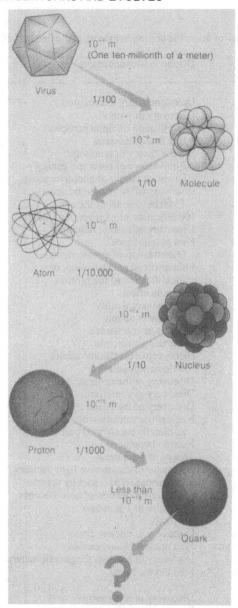

Figure 3.2. Sizes in the structure of matter (from Ref. 61, reprinted by permission). The characteristic sizes decrease from top to bottom of the figure: ~ 10^{-7} m the size of the smallest living organism, the virus; ~ 10^{-9} m the size of the molecule; ~ 10^{-10} m the size of the atom; ~ 10^{-14} m the size of the nucleus; ~ 10^{-15} m the size of the proton; less than 10^{-16} m the size of the electron; and less than 10^{-18} m the size of the quark. These latter distances are the shortest that can be presently measured.

Table 3.9. Examples of scientific accomplishments in the first half of the 20th century[a]

Principal scientist(s)	Discovery/Advancement	Year
M. Planck	Quantum theory of radiation	1900
P. Villard	Discovery of γ-rays	1900
E. Rutherford	Identification of alpha particles as a helium nucleus ·	1903
A. Einstein	Special theory of relativity/ Equivalence of mass and energy/ Quantum theory of photoelectricity	1905
E. Rutherford	Discovery of the atomic nucleus/ Planetary model of the atom	1911
F. Soody	Identification of isotopes and isobars	1911
V. F. Hess	Discovery of cosmic radiation	1912
N. Bohr	First atomic model/ Quantization of atomic states	1913
E. Rutherford	First artificial nuclear reaction/ Identification of the proton	1919
A. H. Compton	Compton effect	1922
L. de Broglie	Particle-wave duality	1924
E. Schroedinger	The wave equation/ Quantum mechanics	1926
W. Heisenberg	Uncertainty principle	1927
H. J. Müller	Discovery of mutations[b] caused by ionizing radiation	1927
J. Chadwick	Discovery of the neutron	1932
H. Urey	Discovery of deuterium	1932
C. D. Anderson	Discovery of the positron[c]	1932
J. D. Cockroft/ E. T. S. Walton	Nuclear transmutation by artificially accelerated particles	1932
W. Pauli	Nutrino hypothesis	1931
W. Pauli/E. Fermi	Theory of beta decay	1933-4
I. Curie/F. Joliot	Discovery of radioactive light nuclides	1934
E. Fermi	Transformation of nuclei by neutron capture/Production of new isotopes	1934
C. D. Anderson/ S. Neddermeyer	Discovery of ± μ meson	1936
O. Hahn/F. Strassmann	Discovery of nuclear fission	1938
E. Fermi and Collaborators	First nuclear chain reaction	1942
Avery/Maclead/McCarty	Discovery of DNA as the genetic material	1944
The Manhattan Project Scientists	First atomic bomb	1945
C. F. Powell	Discovery of ± π meson	1946
E. Teller and Collaborators	First fusion (hydrogen) bomb	1952

[a] Based in part on Refs. 62 and 63.
[b] An inheritable alteration of the genes.
[c] Proposed by P. Dirac in 1931.

made exorbitant peace-time investments in science. To quote Vannevar Bush,[64] "It was through the Second World War that most of us suddenly appreciated for the first time the power of man's concentrated efforts to understand and control the forces of nature. These developments have established that science and technology through large-scale organized research can solve even the most difficult technological problems such as the atomic and hydrogen bombs, radar, nuclear reactors, satellites, and space exploration, and can lead to miraculous new technologies under government sponsorship." In this way the Atomic Energy Laboratories [most of which were later transformed into gigantic Research and Development (R&D) complexes called "National Laboratories" for either defense, or energy, or "all-purpose R&D"] spawned in the USA (and similar ones in the Former Soviet Union and many Western countries) along with a host of other types such as the Federal Laboratories, the National Accelerator Laboratories, and new structures in other fields such as the National Institutes of Health. Military industrial complexes grew bigger, project laboratories were established, "Big Science" emerged, and emerged gigantic at birth. The "Atoms for Peace Program" in the 1950's boosted atomic, nuclear, radiation, and reactor physics research and applications in many countries. In the late 1950's the challenge of the Sputnik led to the superstructures of NASA and opened the door for not just the road to the moon, but, also, the road to what seems to have been the golden decade of science (late 1950's to mid-1960's). Thus, from the end of WWII until the middle 1960's, science, with minor fluctuations, continued to grow rapidly. Money, public faith, government support, prestige, and political influence were abundant. Physics in particular dominated the postwar science. Gradually, however, chemistry, engineering, biology, and life sciences gained in influence and share of resources. National security considerations and the evolution of the cold war greatly influenced science in this period during which great numbers of scientists migrated mainly from Europe to North America -- the famous "brain drain" of Europe.

In the decade of the 1960's the economist and the systems analyst began exercising increasing influence on science policy. This decade also -- slowly but persistently in its opening and violently in its closing years -- saw the beginning of the period of disenchantment with science and technology. As we have discussed in the pervious chapter (see also Chapter 10), scientific research was linked to the war machine and to environmental deterioration. On the other hand, the so-called "environmental crisis" led to the establishment of new environmental agencies and new research laboratories. It was quickly joined -- and to some degree for a while it was overtaken -- by the ensuing "energy crisis" of the 1970's which, in turn, caused the spawning of the "Energy Laboratories" and the reorientation of old research and academic institutions toward energy-related fields. Science was asked to serve the "National Need," whatever the meaning and whatever the need. And as the Cold War was coming to an end in the late 1980's, a part of science's castle collapsed, just like the Berlin Wall. Many scientists and technologists had to switch their activities from military to civilian and many scientific institutions had to reinvent new missions such as cleaning up the many places that had been polluted by fifty years of radioactive and chemical contamination and storing and safeguarding of the dangerous materials that have been produced. That transition is yet to be completed. Each and every one of these "crises" and events influenced considerably the patterns of scientific inquiry and restricted the scientist's freedom, but had not stopped the overall growth of science. The remarkable scientific accomplishments of the post World War II period are too numerous and too well known: the transistor, the laser, the peaceful uses of the atom and its nucleus, the physics of the electron and the associated electronic technologies in computers and communications, the tailoring of molecules (the marvels of chemical synthesis) and materials for the needs of man, the manipulations of the properties of matter through the interplay between light and charge and their use for the emerging fast technologies of the future, the uncovering of the DNA structure and subsequent advances in genetics and bioengineering, the eradication of disease and the miracles of modern medicine, the exploration of space, the Sputnik, the manned space flights, the walking of man on the surface of the moon, the Viking missions to Mars and so on (see, also, Table 3.10 and Chapters 4 and 7).

But perhaps the most distinct scientific development in the post World War II era is the rapid change in the character of science toward "social need." Science has increasingly been forced to orientate a great deal of its effort from understanding to control, from basic to applied, from extending man's knowledge into the endless frontier of the unknown to the utilization of existing knowledge; from the free and unrestricted inquiry of basic research to the constrained areas of "mission oriented" and "project-oriented" research, and the "service tasks" (Chapter 7). It can be said that recently science is experiencing a gradual but profound double shift from physics and chemistry toward the life and biomedical sciences, and from basic to applied and industrial research.

Thus, not only the tasks and the definition of science have been broadened, but also the size and the rate of growth of science have increased. Modern science has become too big, too complex, too elaborate, too expensive, and all too pervasive. It requires large sums of money, large numbers of specialized workers, many and varied kinds of sophisticated equipment and machines(some costing billions of dollars), and gigantic administrative bureaucracies to sustain and manage its functions. It is estimated that 80% to 90 % of all scientists (or science workers, see Chapter 6) who have ever lived are alive today and nearly 90 % of the current stock of scientific knowledge has been discovered within the last half century.

The emergence of Big Science has created many and difficult problems for science itself, for the institutions of science, and for the relation of both to society. *Science has changed because it got too big and it got too big because it was forced to change.* The complexity and the size of modern science, the demands of the scientific/technological problems of modern societies, and the burden of public expectation from science have forced on

Table 3.10. Examples of scientific advances during the latter part of the 20th century

---

- Elementary particles: discovery of the quark and many new particles; advances in understanding the nature of the forces of nature
- Extension of quantum mechanics to complex systems
- Peaceful uses of the atom and its nucleus and their radiations and energy; new sources of energy
- The explosion of the physics of the electron and the associated technologies in communications
- The transistor and its applications
- The laser and its applications
- The tailoring of molecules to the needs of man, man-made substances, chemical synthesis, new drugs, and biotechnology
- Materials science, interplay between light and charge and its use for fast technologies, fiber optics, high temperature superconductors
- Unraveling of the structure of DNA and the discovery of Recombinant DNA
- Advances in genetics, bioengineering and their implications for health, disease, and the ethics and values of man
- Cosmology: the discovery of cosmic microwave background radiation, quasars, pulsars, neutron stars, black holes
- Space exploration: the Sputnik, manned space flights, man's landing on the moon, the Viking and the Galileo missions
- The computer, information handling and compression, communication systems and their impact on society

---

science many and distinct changes (see Chapters 5, 6, and 10). The huge government expenditures and the gains that can be realized by the exploitation of scientific knowledge have made modern science one of the biggest businesses ever. Universities, national laboratories, research institutes, industries, and groups and individuals of all kinds have found scientific research a profitable enterprise and a good way to earn a living. They compete immensely and fiercely for the "research budget" and the research contract. The advancement of scientific knowledge is no longer an academic activity carried out in universities, or the sole purpose and responsibility of the scientific community. *Modern science is nourished, administered, directed, and controlled by many -- mostly government -- bureaucracies.* There are concerns that the maintenance of the standards of excellence rests no longer with the academic institutions and that the scientific effort is undergoing a dangerous drift from excellence to mediocrity. There are concerns also that the scientist has been overtaken by the science worker, and that both have been overrun by the science administrator, the bureaucrat, and the peripheral. There are furthermore concerns that new loyalties and more complacent attitudes are eroding the ethical basis underlying the scientific tradition (see Chapters 6 and 10).

3.8. THE INVALUABLE VALUE OF SCIENCE'S HEREDITARY PAST

Before closing this chapter it is only natural to pose the question: What does the history and the tradition of science teach us? What does the hereditary past of science offer the student in science or the modern person? The history of science may be of little help to the modern scientist in terms of new methods or new discoveries, but there is immense value in its appreciation, guidance, and principles. In science as in the rest of human experience the past is immanent in the present in many and important ways. Some of these are elaborated upon in this section and others are discussed in Chapters 5 and 6.

Every Step and Every Contribution Was an Important Development at the Time it Was Made. Vividly inscribed in science's hereditary past is the embryology and evolution of truth through the ages. The evolution of scientific knowledge is difficult to quantify. Nevertheless, it seems to be underlined by a laborious step-by-step process (Figure 3.3). In the step-by-step-type advancement of science every previous step in discovery, however small, contributes to the emergence of the next. Every step is an important achievement at the time it is made and every piece of discovery paves the way for the next. It is scientifically honest to follow the development of ideas back to colleagues whose names are forgotten, never mentioned or cited any longer. It is important too to recognize the dignity of even the humblest effort in science, for there is virtually no brilliant

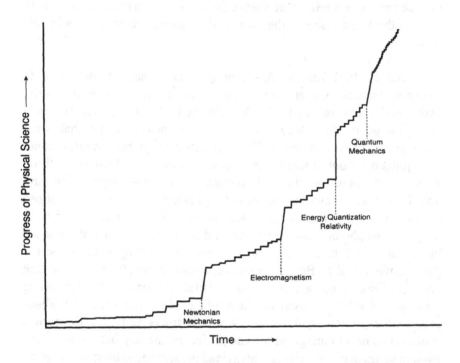

Figure 3.3. Schematic illustration of the step-by-step advancement of physical science. Successive -- and progressively more frequent -- steps lead to new syntheses and constructs responsible for step-function-type (abrupt) advances.

discovery which has not been preceded and which has not been prepared by many minor and obscure endeavors. Scientific discoveries, as we have seen in this chapter, and as others have noted,[65] are seldom single events attributable to single individuals at specific times. To the individual masters whose names we frequently mention in this book belongs the synthesis and the vision of new paradigms, but it is to the body of science that belongs the preparation of the platform which made possible their syntheses and paradigm constructions. This has been the case for Newton's synthesis, for Maxwell's synthesis, for Einstein's propositions, for quantum mechanics, and for the yet to come synthesis of man's knowledge on the forces of nature. All were, and most likely all will be, the elegant synthetic terminations of long, laborious, and tortuous paths feebly illuminated by the work of many. Science's past therefore teaches that there is something in every scientist's work which ultimately honors humanity. Students in science need to recognize that science did not start with their own research, or with the latest issue of the journal they happen to read, or with the computer.

Science's Past Teaches Modesty. In science there are no infallible workers. Aristotle, Copernicus, Kepler, Newton, Einstein, Bohr, Curie, Fermi and every scientist who has devoted his life to science made propositions, or held views, or conducted measurements that were subsequently proven incorrect. This instructs caution, but not refrainment from judgement out of fear of error. Science has been advanced by those who were cautious, but daring in putting forward bold explanations and novel ideas: Newton proposed gravity, Maxwell proposed electromagnetism, Planck proposed quantization of energy, Einstein proposed the photonic nature of light and relativity, Rutherford proposed the existence of the nucleus and the neutron, de Broglie proposed the particle-wave duality, Bohr proposed complementarity, Pauli proposed the neutrino, Dirac proposed the positron, Hubbell proposed the expanding universe, Mendel proposed the laws of heredity, Darwin and Wallace proposed evolution, and so on. New constructs and new theoretical frameworks never emerge whole and perfect in one step but rather a step-by-step series of discoveries and ideas lead to them and subsequently refine them. Ideas and bold propositions often are met with resistance not so much because of malice, but because of intellectual and thought barriers and a natural agony which often accompanies the accommodation of the new within the frame of the existing at the time the ideas are proposed.

Ideas, propositions, and theories once proven become self-evident. And yet, eventually they are superseded by others for science's answers however profound are temporal. Galileo overruled Aristotle, Copernicus overruled Ptolemy, Kepler overruled Copernicus, Newton overruled Aristotle, Einstein overruled Newton, and Bohr won the argument over Einstein's views on quantum mechanics. This is the process of science, a self-correcting and a self-improving process. Constructs change, are modified, or are abandoned. Theories, too, come and go sometimes abruptly.

The Value of Thorough and Accurate Experimental Data and the Power of the Experimental Method and Technique. The accurate observations of Brahe led to Kepler's elliptical orbits, the accurate and thorough measurements on blackbody radiation in Germany led to quantum physics, and the crystallographic data on macromolecular structures led to the DNA double helix. Pasteur's ingenious simple techniques for sterilization -- his swan-necked flasks -- answered the questions of the existence of germs, solved the 2000 year-long problem of spontaneous generation, and led to the development of the sterilization process. Rutherford's use of alpha particles led to the discovery of the nucleus. Chadwick's technique (the use of paraffin wax and hydrogen gas to force the invisible neutron to make itself visible) led to the discovery of the neutron. Classic examples of the way science works. Humble but ingeniously powerful ways of "Little Science" so different from the ways of today's "Big Science."

The Style of the Masters in the Execution of Science. A great deal can be learned about science and its practice from the past practitioners of science (Chapter 6), especially the *Masters* -- people of varied virtue and vice but with commonality in drive, boldness, persistence, intuition, novelty of ideas, pathos for knowledge, conceptual simplicity, and an uncanny ability to relentlessly scrutinize their findings. Those who concentrated on the whole and the search for the deeper general structure of nature deductively seeking general paradigms from axiomatic postulates and reflective thinking such Newton and Einstein. Those who focused on the parts and induction and discovered the facts about nature by coming closer to it and in a somewhat private and confidential manner nature answered their questions revealing itself to them, scientists such as Faraday, Rutherford, Pasteur, and Fermi. Those whose breath and synthetic ability guided science for centuries such as Euclid (geometry/mathematics), Aristotle (general science/physics/biology), Ptolemy (astronomy), Newton (classical mechanics/dynamics), Maxwell (electromagnetism), and Darwin (evolution). Those (e.g., Bohr) who embraced philosophy in order to better understand science. If "imitation of a great man is one form of moral education," as André Maurois maintained, learning by example the execution of science is the corner stone of science education.

Discoveries Come in Many Ways. They come unanticipated by theory and unaided by pre-existing constructs. They are made through the inductive method of experimental science such as the discovery of the photoelectric effect, the discovery of the electron, the discovery of the atomic nucleus and the discovery of fission. They follow the prediction of theory, such as the discovery of the neutrino, the discovery of the positron, and the discoveries of many elementary particles. They are picked up accidentally and serendipitously by the "prepared" scientist, such as the serendipitous discovery of the electric current by Galvani, the generation of magnetism by an electric current by Oersted, the generation of X-rays by Roentgen, and the observation of radioactivity by Becquerel. They are uncovered by deliberate search and skillful ingenuity, such as the discovery of the neutron by Chadwick and the discovery of the structure of DNA by Watson and Crick. They are squeezed out of the grasp of nature by persistent and stubborn thinking guided by experimental facts, such as the discovery of elliptic orbits of the planets by Kepler guided by the accurate observations of Brahe, or the quantization of energy by Planck guided by the blackbody radiation measurements. They emerge from the brutal force of "megascience" experiments, such as the discoveries of transuranium elements, or the many elementary particles made possible by the gigantic accelerators of Big Science. And they still come in other and yet unforeseen ways.

Shrinking Time Between the Announcement of a New Discovery and its Confirmation or Rejection. As science progresses, the time of the announcement of a new discovery and its confirmation, rejection, or acceptance has been narrowed to virtually zero. The long and tortuous paths that lead to answers to fundamental questions are becoming shorter and shorter and at times very short indeed. It takes no longer centuries to challenge the ideas of the Masters or the correctness of an experimental claim as can be seen from the few examples in Table 3.11.

Shrinking Time Between Discovery and Application. The increasing strength of the bond between science and technology is illustrated by the shortening of the time that separates scientific discovery and its application. Examples showing this shrinking of the time lapse between scientific discovery and technological development are given in Table 3.12. In modern times this time lapse has shrunk to almost zero (Chapter 7).

Distinct Characteristics and Principles of Science. Science's past teaches that there are certain distinct characteristics and principles of science that qualify its functions. These distinct characteristics and principles of science are discussed in Chapter 5. They need to be recognized and adhered to if science is to serve well and in a balanced way man and his societies.

Table 3.11. Examples of the shrinking of the time lapse between the origin of an idea or a scientific prediction or an announced discovery and its clarification, confirmation, or rejection

Scientific idea/ Prediction/ Announced discovery	Origin → Clarification/ Confirmation or confirming discovery/ Rejection	Time period (Years)
Atomic constitution of matter	Leucippus/Democritus (5th century, B. C.) → Lavoisier/Dalton (18th century, A. D.)	2300
Shape of planetary orbits	Ancient Greece (5th century, B. C.; circular) → Kepler (17th century, A. D.; elliptic)	2200
Electricity and magnetism	Ancient Greece (6th century, B. C.) → Europe (17th century, A. D.)	2300
Heliocentric theory of the solar system	Aristarhus (3rd century, B. C.) → Copernicus (16th century, A. D.)	1900
Spontaneous generation of life	Aristotle (4th century, B. C.) → Pasteur (19th century, A. D.)	2300
Light is made of particles	Newton (1704) → Einstein (1905)	201
Light is an electromagnetic wave	Maxwell (1864) → Hertz (1887)	23
Wave properties of electrons	de Broglie (1924) → Davisson/Germer (1927)	3
Existence of positrons	Dirac (1931) → Anderson (1932)	1
Existence of neutrino	Pauli (1931) → Reines/Cowan (1956)	25
Nuclear fission	Hahn/Strassmann (1939) → Many groups	Days
The quark	Gell-Mann/Zweig (1964) → Stanford Linear Accelerator Scientists (1967-1973)	3-9
Cold fusion and its rejection	Pons/Fleischmann (1989) → Many groups	Immediately

Table 3.12. Some examples illustrating the shrinking of the time lapse between scientific discovery and technological development (as given by Krazberg[66]) (see also Chapter 7)

Basic scientific discovery	Time lapsed for corresponding technology to be developed (Years)
Steam engine[a]	1700
Principle of photography[b]	200
Electric motor	40
Thermionic effect to triode tube	35
Nuclear power	5
Transistor	5
Transparent plastics	2
Laser light	1.5
Superconducting magnets	1.5

[a] Designed in Alexandria, Egypt.
[b] Designed by Leonardo da Vinci.

3.9. REFERENCES AND NOTES

1. R. Dubois, *Reason Awake*, Columbia University Press, New York, 1970, p. 34.
2. J. Ziman, *The Force of Knowledge*, Cambridge University Press, Cambridge, 1976, p. 57.
3. B. Russell, *The Future of Science*, Philosophical Library, New York, 1959.
4. D. Bohm, quoted by E. Wigner, Foundations of Physics **1**, 35 (1970).
5. J. Bronowski, *Science and Human Values*, Harper & Row, Publishers, Inc., New York, 1965, p. 13.
6. W. Heisenberg, Daedalus LXXXVII (Summer 1958), p. 95.
7. A. N. Whitehead, *Science and the Modern World*, The MacMillan Company, New York, 1926, p. 146.
8. Homer probably made the first recorded observation in radiation chemistry when he commented on the sulphurous smell in the atmosphere following lightning (G. Hughes, *Radiation Chemistry*, Oxford Chemistry Series, Clarendon Press, Oxford, 1973, p. 1). The smell is due to the oxides of nitrogen and ozone generated by reactions initiated by the lightning discharge.
9. L. Motz and J. H. Weaver, *The Story of Physics*, Plenum Press, New York, 1989.
10. E. Segrè, *From Falling Bodies to Radio Waves*, W. H. Freeman and Company, New York, 1984.
11. Ref. 7, p.10.
12. There were a few exceptions such as Archimedes of Syracuse (287-212 B. C.) who invented the lever, Ktesibios of Alexandria (ca 270 B. C.) who founded hydraulics, and

Heron of Alexandria (ca 60 A. D.) who invented dioptra (see, for instance, A. G. Drachmann, in *Technology in Western Civilization*, edited by M. Kranzberg and C. W. Pursell, Jr., Oxford University Press, New York, Vol. 1, 1967).

13. P. Feyerabend, in *Physics and Our View of the World*, J. Hilgevoord (Ed.), Cambridge University Press, Cambridge, 1994, p. 143.

14. Plutarch's Moralia, Vol. II, pp. 384-389, F. C. Babbitt (translation), W. Heinemann, Ltd., London, 1928, pp. 382-389.

15. Physics Today, March 1996, p. 53.

16. Two notable events in this period (middle of the 14th century to the beginning of the 17th century) are the invention of printing attributed to Johann Gutenberg of Mainz, Germany, in the 1440's, and the work of Leonardo da Vinci in Italy.

17. Ref. 9, pp. 27 and 28.

18. The ratio of the distance, x, a body moves in a given time, divided by the time, t, it takes to move the body through that distance, is the average speed of the body, $<\upsilon> = x/t$. The body's motion, however, has a direction and thus not only its average speed is significant but also its direction. A body's motion is thus more accurately described by another quantity, its velocity, which is a vector quantity, an arrow that points to the direction of the motion at each time and its magnitude is the body's speed. A body's velocity at any instant of time is called the instantaneous velocity of the body and is represented as $\upsilon = dx/dt$. Similarly, the average acceleration in a time interval $\Delta t = t_2 - t_1$ during which the velocity changes by $\Delta \upsilon = \upsilon_2 - \upsilon_1$, is equal to $\Delta \upsilon / \Delta t$. The instantaneous acceleration a is the limiting value of the average acceleration as Δt approaches zero, i.e., $d\upsilon/dt$, which is the derivative of υ with respect to t (The boldface letters denote vector quantities.).

19. Equation (1) can be expressed in terms of the particle's linear momentum $p = m\upsilon$. The net external force, F, equals the time-rate-of-change of p

$$F = d/dt(p) = d/dt(m\upsilon) = (\upsilon)dm/dt + md(\upsilon)/dt = ma, \text{ only when } dm/dt = 0.$$

Newton's law applies only when the body's mass is not changing with time. Such a change was shown to occur 200 hundred years later when the speed of the body approaches the speed of light (Section 3.5).

20. Gottfried Wilhelm von Leibnitz also arrived at the calculus independently.

21. C. P. Snow, *The Two Cultures: and a Second Look*, Cambridge at the University Press, 1965, p. 29.

22. R. Dubois, *Reason Awake*, Columbia University Press, New York, 1970, p. 85.

23. See for instance Chapter 4 of Ref. 10.

24. Whereas positive and negative charges can exist independently of each other, individual magnetic poles cannot. This difference produces a remarkable asymmetry between electricity and magnetism even though the two phenomena are intimately related.

25. D. C. Giancoli, *Physics for Scientists and Engineers*, Second edition, Prentice Hall, Englewood, N. J., 1988.

26. J. C. Maxwell, *A Treatise on Electricity and Magnetism*, Oxford, Clarendon Press, 1873; Third Edition, Dover, New York, 1954. See also Ref. 25, p. 729.

27. S. Glasstone, *Textbook of Physical Chemistry*, Second edition, MacMillan and Co. Limited, London, 1956, p. 191.

28. L. G. Christophorou, E. Illenberger, and W. Schmidt (Eds.), *Linking the Gaseous and*

the Condensed Phases of Matter, Plenum Press, New York, 1994.

29. There were a number of prominent scientists who dismissed the model of the atomic constitution of matter foremost among them Friedrich Wilhelm Ostwald.

30. See for instance, R. J. Puddephatt, *The Periodic Table of the Elements,* Oxford Chemistry Series, Clarendon Press, Oxford, 1972.

31. The Periodic Table of the Elements is built up according to basic principles of atomic theory. The modern periodic classification of the elements depends on atomic structure and the energy levels of electrons in atoms arranged in terms of quantum numbers and specific rules. Each element in the Periodic Table contains one more nuclear charge than the preceding element. This additional charge is neutralized by addition of an electron which is put into the lowest energy orbital available in accordance with certain rules and in accordance with the Pauli Exclusion Principle. The elements can then be classified into periods depending on which electron shell is being filled. Each period begins with the occupation of an *ns* orbital and ends when the *np* orbitals are full.

32. R. Morris, *Time's Arrows,* Simon and Schuster, New York, 1985, Chapter 7.

33. R. P. Feynman, *The Character of the Physical Law,* The M.I.T. Press, 1967.

34. H. Price, *Time's Arrow and Archimedes' Point,* Oxford University Press, Oxford, 1996.

35. The macroscopic, thermodynamic description generally deals with averages and the probabilistic elements introduced by quantum mechanics play no role. Therefore, defining the arrow of time this way seems inapplicable for the microcosmos. There are no laws of physics that make backward-in-time-motion impossible. In fact, in 1949 Richard Feynman suggested that in certain cases such reversed motion might be observed and might explain the behavior of antiparticles. For instance, in this view a positron is an electron that moves backward in time; a positron moving forward in time and an electron moving backward in time have the same properties. The paradoxical possibility of time-reversed-motion is a consequence of the fact that on the subatomic level there is generally no arrow of time.

36. See, for example, a recent book by I. Prigogine (*From Being to Becoming,* W. H. Freeman and Company, San Francisco, 1980) discussing the relation between the physics of *"being"* and the physics of *"becoming."*

37. E. Segrè, *From X- Rays to Quarks,* W. H. Freeman and Company, New York, 1980, pp. 65 and 66.

38. Ref. 37, p. 76.

39. According to A. Pais (Science **218,** 17 December 1982, p. 1193), Einstein "deprecated the idea that relativity is revolutionary and stressed that his theory was the natural completion of the work of Faraday, Maxwell, and Lorentz."

40. A reference frame is a set of coordinate axes fixed to some body (or group of bodies), such as the earth or the moon.

41. D. Lindley, *The End of Physics,* Basic Books, New York, New York, 1993, p. 59.

42. E. P. Wigner, Physics Today, March 1964, p. 34.

43. Ref. 41, p. 87.

44. A number of tests of the predictions of the general theory of relativity are discussed in J. D. Barrow and J. Silk, *The Left Hand of Creation,* Oxford University Press, New York, 1983

45. This phrase is the title of a book by F. Close, *The Cosmic Onion,* American Institute of Physics, New York, 1986.

46. According to Pippard [A. B. Pippard, in *Electron*, M. Springford (Ed.), Cambridge University Press, Cambridge, 1997, p. 1] the word *electron* was introduced in 1891 by Johnstone Stoney.

47. According to Faraday's two basic laws of electrolysis the amount of material which is deposited on an electrode is proportional to the quantity of electricity which has passed through the solution and to the atomic weight of the material. These findings suggested that the quantity of electricity which is required to deposit a molecular weight of material (made of singly charged carriers) on an electrode is $Q_{mole} = N_A e$. The amount Q_{mole} is known as the Faraday and may be accurately measured. It is the product of Avogadro's number (6.022×10^{23}) and the smallest electric charge e (1.602×10^{-19} C) and represents the electric charge carried by one mole of singly ionized species. Moreover, from the Faraday the ratio e/M of the fundamental charge e to the mass M of an atom can be measured.

48. Gas discharge studies themselves were made possible by advances in vacuum techniques which allowed pressures well below atmospheric and made gas-discharge studies easier because gases break down at relatively lower voltages when at lower pressure. An early indication of the electrical constitution of matter came from gas discharges when Hertz observed that when either of the two spheres in an electrical oscillator was irradiated by ultraviolet light, the gas discharge occurred more quickly than without light irradiation: the light ejected electrons from the irradiated sphere which triggered the discharge. In essence this was the first observation of the photoelectric effect.

49. See Appendix B for the currently accepted values of e and m.

50. We shall see later that such disintegrations can be externally induced (artificial nuclear disintegrations).

51. F. Close, *The Cosmic Onion*, American Institute of Physics, New York, 1986, p. 14.

52. L. G. Christophorou (Ed.), *Electron-Molecule Interactions and Their Applications*, Academic Press, New York, Volume 1, 1984, p. 559.

53. H. E. White, Phys. Rev. **37**, 1416 (1931).

54. V. F. Weisskopf, Science **207**, 14 March 1980, p. 1163.

55. A. Pais, Science **218**, 17 December 1982, p.1193.

56. Ref. 37, pp. 182 and 183.

57. The various elements have different isotopes. For instance, hydrogen usually has one proton and no neutrons, but a small percentage of the hydrogen atoms have one or two additional neutrons. The hydrogen atom with one more neutron than the usual atom is the "heavy hydrogen" or the deuterium isotope. The hydrogen with two more neutrons is the isotope tritium. The total number of neutrons and protons is placed as a subscript to the atomic symbol to distinguish the various isotopes. The number of protons is shown in the superscript. The net numbers of protons and neutrons are separately conserved in nuclear reactions.

58. Ref. 51, p. 25.

59. This effort of producing new elements has continued unabated since. The transuranium element with $Z = 110$ has recently been reported.

60. We follow the account given by Segrè (Ref. 37, p. 209).

61. *To the Heart of Matter-The Superconducting Super Collider*, Universities Research Association, Washington, D.C.

62. A. Pais, *Inward Bound*, Clarendon Press, Oxford, 1986.

63. R. D. Evans, *The Atomic Nucleus*, McGraw-Hill book Company, Inc., New York, 1955.

64. V. Bush, *Science Is Not Enough*, William Morrow & Company, Inc., New York, 1967.
65. Thomas Kuhn, *The Structure of Scientific Revolutions*, Revised edition, Chicago University Press, Chicago, IL, 1970.
66. M. Kranzberg, American Scientist **55**, 48 (1967).

CHAPTER 4

Science: the Penetrator
of the Physical Universe

If the first half of the 20th century has been a period of revolutionary scientific ideas, its second half has been a period of scientific diversity and increased complexity. During the second half of the 20th century science penetrated deeper into the intricate texture and dimensions of the cosmos, the finely weaved constitution and structure of matter, and the fabric of life. Its most fundamental elements during this period are based on relativity, quantum mechanics, and -- in life sciences -- DNA (deoxyribonucleic acid). Most of its successes are a consequence of powerful new experimental methods and novel instruments small and big alike.

4.1. THE MINUSCULE AND THE GIGANTIC MICROSCOPES OF MODERN SCIENCE

The power of the experimental instruments of modern science will be illustrated by two examples. One from Little Science: a minuscule instrument, the scanning tunneling microscope. The other from Big Science: a gigantic machine, the particle collider. The latter is, in essence, like the former, a microscope. Through both types of instruments the scientist is able to "see" hitherto inaccessible regions of the cosmos.

4.1.1. Scanning Probe Microscopes

Microscopes have played a fundamental role in the development of experimental science ever since microscopy began in the 15th century. The optical microscope was developed in late 17th century and has been an integral part of experimental science since. Two of its distinct early successes were the revelation of the existence of single cells and bacteria. However, optical microscopy is limited in one fundamental way: it cannot resolve atomic-size structures because the wavelength of the visible light it uses is very much larger than the diameter of the atoms it looks at. The realization that electrons of sufficient kinetic energy (in the keV range) have wavelengths comparable to atomic dimensions, allowed science to overcome this shortcoming by developing in the 1930's the electron microscope which can resolve nanometer-scale structures. The electron microscope, however, is not capable of resolving the structure of surfaces because the energetic electrons it uses penetrate deep into matter revealing little of the surface, and surfaces are important to science and technology. A major advance in this direction came in the 1960's with the development of the atom-probe field-ion microscope which allows single atom imaging, single atom chemical analysis, and studies of the dynamical behavior and energetics of atoms on metal surfaces.[1] However, this microscope, too, has its limitations: it requires the sample to be located on a fine tip a few angstroms wide and to be stable under the high fields associated with this geometry. Imaging of individual surface atoms with unprecedented resolution[2-4] became a reality in 1981 with the invention of the scanning tunneling microscope (STM). The scanning tunneling microscope is a minuscule, "pocket-type," instrument which needs no lenses to form images because it uses no free particles (photons or electrons) like the optical or the electron microscopes. The band electrons in the sample under study are the radiation source.

The principle and basic elements of the scanning tunneling microscope are schematically shown[5] in Figure 4.1. The system works best under ultra-high vacuum conditions and for conducting surfaces. The tip of an extremely sharp needle is positioned so close to the surface of the sample that the wavefunctions of the electrons in the tip overlap the wavefunctions of the electrons in the surface. When a small voltage is applied between the probe and the surface, electrons tunnel through the vacuum space between the surface and the probe. Because the electron wavefunctions decrease exponentially with distance, the tunneling current is extremely sensitive to

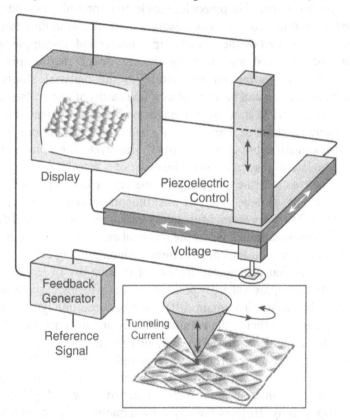

Figure 4.1. The scanning tunneling microscope senses the atomic topography of a surface by means of the current tunneling across the gap between the tip of a probe and the surface. A voltage is applied to the tip which is moved toward the surface (which is conducting or semiconducting) until a tunneling current begins to flow. The tip is then swept across the surface and a feedback mechanism senses the tunneling current and keeps the tip at a constant height above the surface. This way, the tip follows the contours of the surface atoms. The motion of the tip is read and processed by a computer. Sweeping the tip through a pattern of parallel lines, a three-dimensional image of the surface is obtained. Since the tunneling current depends both on the tunnel distance and on the electronic structure of the surface, it gives information not only on the surface topography but also on its constitution (from Ref. 5).

the separation of the needle tip and the surface. Consequently, when the probe is scanned across the sample at a constant height above the surface, features of the surface as small as atoms show up as variations in the tunneling current. The translation and distance determining elements of the STM are piezoelectric slaps. The z-piezoelectric slap is used (Figure 4.1) to adjust the tip-surface spacing and the x- and y-piezoelectric slaps to provide the lateral scan. The piezoelectric elements provide position control adjusted to within 10^{-4} Å. The resolution of the scanning tunneling microscope is limited by the smallest tip diameter and tip-sample spacing that can be achieved and also by the nature of the sample under investigation. Depending on the nature of the material studied, the horizontal resolution can be as good as a few tenths of one Å. Besides the operation of the STM in a vacuum environment as just described, it can also be operated at atmospheric pressure and even under water making its use in biological applications feasible. However, for non-conducting surfaces the lateral resolution is of the order of 5 Å to 10 Å.

The success of the scanning tunneling microscope led to the development of a new class of instruments called *scanning probe microscopes*. These are atomic-resolution instruments which yield detailed information on an atomic level of the physical and electronic properties of surfaces, and image biological structures including living cells.[6] They are all based on similar principles. Instead of making use of lenses to form images they all employ a sharply pointed sensor tip to detect some property of the sample surface. The main difference between one type of a probe microscope and another is the nature of the tip and the corresponding tip-sample interaction. Like the STM, they rely on the precise control of the movement of either the probe or the sample in three dimensions by piezoelectric elements as a means to obtain probe-to-surface distances that result in physical interactions which are transducible to high resolution surface topography. Besides the scanning tunneling microscope, another prominent member of this class of new instruments is the atomic force microscope (AFM). While the STM exploits electron tunneling phenomena (which are favored to occur when a conductive probe is placed in very close proximity to a conductive surface separated by only a thin insulating gap), the AFM is based on the attractive/repulsive forces occurring between a cantilever-mounted probe and a sample and is not restricted to conductive probe/sample systems. The ability of the atomic force microscope to resolve with high resolution surface structures of non-conductive specimens and to operate in aqueous solutions makes it a promising instrument for determining the structures of biological molecules under native conditions.[5,6]

Scanning probe microscopes are just one example of the many ingenious ways modern science uses to penetrate diverse and complex areas of nature.

4.1.2. Particle Colliders

Particle colliders are one of the most powerful, precise, and complex instruments of modern science. They represent the ultimate "microscopes" for looking into the very heart of matter. Miraculously, the knowledge they provide on the infinitesimally small aids man's understanding of the very big, the universe and its creation. It is as though they have united the concepts of the microscope and the telescope into one!

The genealogy of today's mammoth machines traces to the humble vacuum tubes, the earliest accelerators in which electrons were speeded up by the voltage difference between two oppositely charged electrodes. These were followed by the Cockroft-Walton and the van de Graff machines, the modern linear accelerators, the synchrotrons, and the colliders. The last three are awesome experimental machines each an improvement over its predecessor. Thus, the limitations of the straight-line linear accelerators with regard to the energy to which particles can be accelerated by the finite length of their flight path, is overcome in the cyclotrons where the particles are magnetically guided along spiral paths and are accelerated many cycles by a single electric field, achieving this way much higher energies. Today's synchrotrons allow particles (for instance electrons and protons) to gain huge energies by passing millions of times through the electric fields that accelerate them, and today's colliders by allowing two carefully controlled particle beams to collide with each other at a chosen point in the detectors, overcome the limitation of the fixed-target accelerators where a single particle beam hits a stationary target. The collider has another distinct advantage over the fixed-target accelerator: in the former the two colliding particles have no net motion and thus all their energy becomes available for new reactions while in the latter most of the energy of the projectile particle is wasted to kinetic energy of the debris.

Several colliders exist today. In Figure 4.2 is shown the Tevatron, the proton synchrotron at the Fermi National Accelerator Laboratory (FERMILAB) in Batavia, near Chicago. The protons and antiprotons collide at a combined energy of 1.8 TeV inside the accelerator's two

Figure 4.2. The Tevatron, the proton synchrotron collider at the Fermi National Accelerator Laboratory (FERMILAB) in Batavia, near Chicago (Courtesy of FERMILAB).

massive detectors and are guided by superconducting magnets around a 6.3-km circular path. Had the Superconducting Super Collider (SSC) been built in the United States, it would have provided a proton-proton collision total energy of 40 TeV and its main ring would have had a circumference of 87.1 km.[7] Such gigantic machines are expected to be built in the future as the search for new elementary particles continues.

Indeed, particle physics instrumentation has come a long way since Rutherford's alpha-particle microscope. The accelerators that revealed protons and neutrons within the atomic nucleus and then quarks within the protons and the neutrons, are now reaching the stage of being able to accelerate particles (for instance, electrons and protons and their antiparticles) to energies high enough to look deeper still into the heart of matter, perhaps inside the quarks themselves. They supply to the known stable particles the necessary energy to build up more complex, heavier states, via endothermic reactions and allow answers to persistent questions such as those about the elementarity of matter and the nature of the forces of nature.

The awesome size of today's big colliders makes the minuscule instruments we described in the previous paragraph inconsequentially small. Neither the energy of the electrons in the STM is a match for the energies of the particle beams of the colliders, nor is the sensing system of the STM a measure of the collider's detectors. Nonetheless, both are man's new windows to hitherto inaccessible regions of the cosmos.

4.2. THE DIMENSIONS OF MODERN SCIENCE'S DIVERSITY

4.2.1. The Eternal Quest for the Elementary Constitution of Matter

Man's old notion that all matter has some common ultimate structure continues unabated. Modern science is still seeking nature's elusive elementarity. In the 19th century it was believed that the chemical elements are the ultimate particles. This belief had been abandoned when atoms were found to comprise of electrons and an atomic nucleus. While the electron is still thought to be a fundamental particle (a member of the lepton family), the nucleus is not. That the nucleus is not elemental was clearly shown by the discovery of natural radioactivity. Elements are not immutable because their nuclei were found to transmute from one type to another. In time, the nuclear constituents -- the proton and the neutron -- were similarly discovered to be splittable. They, too, were shown to have no individuality but to transmute from one to the other. In turn, neutrons and protons were themselves found to be made of yet smaller building blocks of matter, the quarks. The quarks are among the elementary particles of which matter is composed of. They play a crucial role in subnuclear spectra, similar to that played by electrons in atomic, and by neutrons and protons in nuclear, spectra. However, although the electrons and the positive ions which are produced when atoms are ionized -- and the protons and neutrons which are produced in nuclear collisions -- are individually detected, individual quarks isolated from one another are not seen.[8] On the existence of quarks rests today's dominant theory of particle physics, the Standard Model. According to this model, all matter is composed of elementary quarks and leptons and the forces between these particles are carried by gauge bosons such as the photon and the gluons.[8] The quarks and the leptons are taken as structureless point-like objects. Are these, then, the true atomic elements of Leucippus and Democritus? Doubtful. Indeed, string theory considers the basic units of matter to be little loops referred to as "strings" rather than point-like particles, and bootstrap theory considers everything to be infinitely divisible,[9] just as Anaxagoras maintained 2500 years ago in ancient Greece.

Science has sought and continues to seek an understanding of the interactions between the basic units of matter and antimatter[10] and the reactions for the existence of the particles themselves. A principal aim of this research is the description of the forces between the elementary particles. What are the forces of nature and what is their origin? How do they affect the fundamental particles and why are particles influenced by some forces and not by others? How are they related? As will be briefly discussed in the next section, it is presently supposed that the interactions between the basic constituents of matter are mediated by a variety of other particles, the so-called "exchange particles." The exchange mechanism can be viewed as the creation or emission of the exchange particles by one unit of matter and its subsequent annihilation or absorption by another unit of matter.

4.2.2. The Nature of the Forces of Nature

Always there is an aura of mystery about nature's plethora of forces. Man's knowledge about the forces of nature advanced substantially during this century. Indeed, much of today's science is directed toward understanding the forces of nature, their sources, their relation, and their ultimate unification. This has been a long and as yet incomplete endeavor. It began with Newton and Maxwell and continues still.

As we have noted earlier, man discovered the gravitational force first. In the 19th century he added to his knowledge of forces the electromagnetic force. In the 20th century he discovered the forces in the microcosmos: the so-called strong and weak forces. Today science accepts four fundamental forces: the strong force, the electromagnetic force, the weak force, and the gravitational force. The relative strengths, effective ranges, mediating particles, and manifestations in the universe of these four forces are listed[11] in Table 4.1. At the small scale of the atomic nucleus, the strong and the weak forces dominate and the gravitational force is extremely weak.

Table 4.1. The forces of nature, their mediating particles, range, and relative strength (based on Renteln[11])

Force	Particles affected	Manifestation	Field quanta	Range (m)	Relative strength
Strong	Quarks	Nuclear power	Gluons	10^{-15}	1
Electro-magnetic	Charged particle	Chemistry	Photons	Unlimited	$1/137$[a]
Weak	Quarks and leptons	Some radioactive decay processes	W and Z bosons	$<10^{-15}$	10^{-5}
Gravity	All particles	Large-scale structure of the universe	Gravitons?	Unlimited	10^{-40}

[a] Strength by which electrons couple to electromagnetic radiation.

The *strong force* is powerful but has a very short-range ($< 10^{-15}$ m). It acts only within the nucleus. It binds neutrons and protons together in the nucleus and quarks together inside protons and neutrons. The strong force is described by quantum chromodynamics, a theory which postulates the exchange of massless particles called gluons.[12] A manifestation of the strong force is nuclear power.

The *electromagnetic force* governs the interactions of electrically charged particles. It binds the negatively charged electrons to the positive charges of the atomic nucleus, it binds atoms in molecules, and it aggregates molecules into bigger structures. In contrast to gravity, the electromagnetic force can be attractive or repulsive. It is much more powerful than gravity (Table 4.1). The electromagnetic interaction of particles is considered to arise from the passage of electromagnetic waves between them, the photons.

The *weak force* is manifested in some radioactive decay processes. It is responsible for the beta decay in which one of the protons in the nucleus transforms into a neutron and a neutron into a proton. When the weak force operates in this way it is very much weaker than the electromagnetic force (by over a factor of 10^8). At high energies it acts in a similar way to the electromagnetic force and at very high energies the two forces appear to be intimately related. The weak force is mediated by particles called W and Z bosons.[13]

Gravity is the most familiar of the fundamental forces of nature. It acts attractively between all particles. It is the force that holds the universe together. It dominates the large-scale structure of the cosmos, the formation of the planets, the solar systems, and the galaxies. It has been referred to as the greediest of all forces[14] because it is always attractive pulling everything in the universe toward everything else. It is the only known force that does not saturate. Its pull steadily increases as more and more matter is brought together. On the microscopic scale the effects of gravity are negligibly small. Gravity is not important as far as the description of the known atomic and nuclear phenomena are concerned. The hypothetical particles which mediate gravity are called *gravitons*. They remain undiscovered.

These four forces -- the strong force, the electromagnetic force, the weak force, and gravity -- control the behavior of all matter and of all phenomena -- microcosmic, macrocosmic, physical, chemical, biological. But how are they related? Science began answering this question, in essence, since Maxwell and as yet has no complete answer. Maxwell's theory unified two previously distinct forces, the electric and magnetic forces which convey the forces between charged objects and between magnetic objects, and established the existence of a single electromagnetic force. In the 1970's Sheldon Glashow, Steven Weinberg, and Abdus Salam showed that the electromagnetic force and the weak force are different manifestations of a single unifying interaction -- the electroweak interaction -- and predicted the existence of the W and Z bosons as the mediating particles of this force. The discovery of these particles in 1983 was an important step toward the unification of three (electromagnetic, weak, and strong) of the four forces of nature. The current understanding of the strong and electroweak interactions is known as the Standard Model of high energy physics. In this theory, as we mentioned earlier, the strong force is carried by gluons and it keeps quarks together inside protons, neutrons, and many other composite particles. The particles which are not made of quarks are called leptons and they do not feel the strong interaction. The electromagnetic and weak forces are mediated, respectively, by photons, and by W and Z bosons. While the theory accounts for the current knowledge on the behavior of elementary particles there are still open questions. A basic problem is that the photon has no rest mass while the W and Z bosons are heavy. This mass difference is the reason the two forces are quite distinct at low energies and show similarities only in the behavior of particles at very high energies. It is related to the profound question of the origin of mass. Experiments with colliders at higher energies may provide answers to these questions.

And there still remains gravity. The electric force and gravity have not as yet been shown to be different aspects of the same entity. The electric force is over 10^{40} times more powerful than gravity. How can forces so different in strength have the same origin? How does gravity look on a small scale? How can general relativity and quantum mechanics be reconciled in a self-consistent theory of quantum gravity? These are questions for science today. The goal of quantum gravity is to understand the gravitational force in the same way the other forces of nature are understood. This is enormously difficult because quantum gravity concerns events at a scale far smaller than any realm yet explored by experimental physics. It is only when particles approach to within about 10^{-35} meters that their gravitational interactions have to be described in the quantum mechanical terms.[11] This distance is 10^{24} times smaller than the diameter of the atom. Too small a distance to probe indeed! It should be noted however that there is evidence that quantum gravity plays a role in black holes and may be elsewhere if the macrocosmos.

4.2.3. Tuning to the Macrocosmic Past

Science's outward look penetrated farther into space and time. Science tuned to new signals the awesome cosmos had sent a long time ago. For example, high-energy astronomy,[14] through principally the X-ray region of the electromagnetic spectrum, has revealed new macrocosmic objects such as neutron stars and new processes such as the diffuse X-ray background radiation[15] which are hidden from the "eyes" of optical telescopes. With the aid of physics, cosmology penetrated time to its beginnings, probed the surface of planets, and traced the universe's evolutionary past. Satellite probes reached Vinus and Mars and man himself landed on the Moon. This immense richness of new knowledge about the constitution, dynamics, and evolution of the cosmos is illustrated below by a few examples.

The Universe is Expanding and Has a History. For centuries the universe was regarded as vast and unchanging. Two discoveries changed this view. The first discovery was made in 1929 by Edwin Hubble who measured a reddening in the light from distant galaxies, a signal that the entire cosmos is in a stage of dynamic expansion. The distant galaxies are receding from each other at a velocity that increases linearly with their distance apart.[16] The universe is expanding in all directions and it must thus have been much denser in the past. The second relevant discovery was made in 1965 by Arno Penzias and Robert Wilson who found a cosmic microwave background (blackbody) radiation corresponding to a temperature of 2.7 degrees above absolute zero whose intensity is the same in all directions in space. The extreme isotropy of the intensity of this radiation indicates that it has to do with the entire universe and that it might have originated in the depths of the universe when the dominant constituent of the universe was blackbody radiation. As the universe expanded thermal radiation cooled down. Both of these discoveries enabled man to follow the expanding universe back in time to its (presumed) origin, to that moment in time when the entire universe must have been reduced to a single point of unimaginable density and was incredibly hot. This instant is called *singularity* "because the simple cosmological models cannot describe what lies before it."[17] It is believed that the cataclysmic beginning of the cosmic expansion had its origin in a primordial explosion -- the Big Bang -- which had occurred some 20-30 billion years ago. Thus, there is scientific evidence that the cosmos has a beginning. Hence, the cosmos has a history. It began, it changes, and it cools as it expands since. An arrow of time points away from the Big Bang.

If the universe is traced backward in time, periods are encountered during which the universe was too hot for neither stars, nor molecules, nor atoms, nor even nuclei to exist. The early universe was radiation dominated. In the beginning there was only the primordial radiation of elementary particles. In time from being a fully ionized plasma the universe turned electrically neutral to a large extent. Nucleosynthesis of the lighter elements occurred[18] when the universe was hot at temperatures of 10^8 to 10^9 K producing the abundances of ^{2}H, ^{3}He, ^{4}He, and ^{7}Li now observed. Today[18] the observed universe expands at a rate of 50 km/s/Mpc, contains a matter density of $\sim 10^{-30}$ g/cm^3, and has a radiation density of $\sim 10^{-33}$ g/cm^3. At the present time matter dominates radiation by a factor of about 10^3.

The Cosmos is Sprinkled with Regions of Extremely Dense and Unique Forms of Matter. Two such regions are those known as *quasars* and *neutron stars*. Quasars were discovered in 1963 and are violently active extra galactic objects believed to be a gravitational concentration of enormous masses of stars and stellar debris in the central regions.[19] They are powerful radio, infrared, and X-ray energy sources and possibly the most energetic objects in the universe. Although no larger than the solar system, they emit more energy than an entire galaxy of one hundred billion stars.[20] Similarly, a neutron star is "essentially a giant nucleus of about 10^{57} neutrons tightly packed in a volume only 10-20 km in radius and with a density of about a billion tons per cubic inch!"[14]

Regions of the Cosmos where Gravity Collapses. Such regions of space in which matter so-to-speak "has been crushed completely out of existence,"[21] and collapsed under the strength of gravity, are called *black holes*. Because of its immense gravitational field, theoretically nothing, not even light, can escape from the surface of this hole in space. Inside the boundary surface of the black hole, the presence of the singularity renders the known laws of physics inoperable.[22] Here is how Shapiro and Teukolsky[23] picture the generation of a black hole: "Consider a collection of particles sprinkled randomly throughout space. Some of the particles will be drawn together by their mutual attraction, creating a local aggregation; the greater density in this region generates a more intense gravitational field, which thereupon sucks in more matter, further strengthening the field, and attracting still more particles. It is disturbing to realize that nothing in the theory of gravitation can stop this process; if the particles interact with one another solely through gravitational effects, they will continue imploding without limit. The density of matter and the strength of the gravitational field will eventually become infinite. A collapsed region of this kind is called a *singularity*, and it signals a failure of the theory."

Persistent Unanswered Questions. We have said repeatedly in this book that it is the nature of science to answer and to raise questions. Some of these questions are amenable to scientific answers, others persist seemingly transcending at least at the moment scientific inquiry. In the latter category are included such questions as: Where did the matter of the cosmos come from? Is there a unique reversibility to the cosmic expansion? What is the ultimate fate of the universe? What is the meaning of singularities? And to these other similar questions can be added. For instance, since science has shown the absence of antimatter in particles associated with low-energy cosmic rays in our Galaxy leads one to conclude that the galaxy is all matter.[18,24] The solar system contains no anti-planets. Why then is there a matter-antimatter asymmetry in the cosmos? Why do we never see antimatter in space? Furthermore, if at least 90 % of the mass of the universe is in some non-luminous form[25] why is there so much dark matter in the universe? Presently unanswered questions that challenge and fascinate the human mind.

4.2.4. Understanding and Manipulating Everyday Matter

Science has also spread and diversified horizontally. From inaccessible places like the nucleus of the atom and the neutron star, science turned as well to the discovery and understanding of the phenomena of ordinary matter, that is, matter such as that encountered in everyday life. Science turned to the investigation of phenomena where the questions are posed in terms of interactions between electrons, atoms, and molecules; in terms of chemical bonds and atomic and molecular theories; in terms of the ways in which atoms fill up space; and in terms of the behavior of nature resulting from new ways atoms rearrange themselves and aggregate. In this, science turned toward complexity and since complexity needs an energy source and a medium, energy and the transformations and reactions it induces along with the role of the medium in building up elaborate structures have been part of the horizontal activity of science. With regard to energy, life makes use of only a small part of the total energy found in the universe and the energy life uses is in its less concentrated forms. With regard to the medium, it is needed for the building up of elaborate structures so that many-body reactions can take place within short times in comparison with the disintegration lifetimes of the more complex structures formed.

Condensed matter science provides ample examples of the penetration of science into the diversity and physical characteristics of everyday matter. Quantum mechanics and sophisticated new instruments and methods enabled science to show how materials with new properties can be synthesized. Materials in which "the mechanical strength, electrical conductivity, optical and dielectric constants, and so on are determined by controlling defects, impurities, composition, and structure."[26] "We have learned in microscopic detail about the structural defects that make metals ductile, the energy-band structure that makes gold gold and glass transparent."[26] Condensed matter science produced dislocation-free silicon for semiconductor processing, developed molecular beam epitaxy, and grew multilayered semiconductor structures having energy gaps that vary periodically in one dimension such as GaAs-GaAlAs and GaSb-InAs. High-purity crystals led to the invention of the transistor. The application of radio frequency and microwave generation and detection methods to solids provided a wealth of information on solid-state resonances (nuclear magnetic, electron paramagnetic, cyclotron, ferromagnetic, and antiferromagnetic), energy levels, crystal field splittings, and relaxation processes. Furthermore, the availability of means to reach low temperatures (down to 10^{-3} K), high pressures (to 10^5 and 10^6 atmospheres), and large magnetic fields (to 20 tesla), allowed studies of condensed-matter systems under extreme conditions and thus the discovery of new phenomena and properties of the solid state of matter.

Two notable examples of recently discovered material structures are worth pointing out: high-temperature superconductors and the fullerenes. The high-temperature superconductors (above the boiling point of nitrogen, 77.2 K) although made of simple layers rely on complex bonding and charge transfer for their stability and intriguing properties. The newly unraveled structures of the fullerenes represent yet another example of the surprising ways atoms (in the case of the fullerenes carbon atoms) can fill space and of the resulting unexpected behavior.

4.2.5. The Brave New World of Molecular Biology

Molecular biology is the science that has revealed to us the gene. Its principal paradigm is DNA. The DNA story is still unfolding, but clearly four steps are evident. First, the discovery in 1944 that DNA is the genetic substance. Second, the discovery in 1953 that DNA has a double helical structure. Third, the discovery in 1973 of recombinant DNA. Fourth, the emerging new field of genetic engineering.

In molecular biology are fused two of the most active fields of modern science: biochemistry and genetics. This is also a field of science where many of the recent advances have resulted from the development of new instruments in other branches of science: microscopes, X-ray crystallography and chromatography; nuclear magnetic resonance, radioactive isotopes, and radioactively labeled compounds; physical probes of three-dimensional structures of macromolecules; ultracentrifuges for separation of cellular particles. The industrial synthesis of biochemicals and antibiotics made it easy to culture or to plant cells free of contamination by faster-growing bacteria and allowed faster and more economic ways to study the genetic and regulatory properties of these cells and of infecting viruses.

The New Paradigm. Based on the work of crystallographers, the physicists James Watson and Francis Crick proposed in 1953 that the structure of DNA is a double helix. Two long thin polymeric chains, two strands of linked atoms twisted around each other forming a long thin rope and clusters of atoms attached to each strand at short regular intervals (Figure 4.3).[27] The bispiral chains of the DNA molecules revealed that the macromolecules which transmit hereditary characters from one organism to the next have a symmetrical structure. Nature "writes its instructions in a four-'letter' alphabet in which each letter is a specific molecule called a nucleotide, and the letters spelling out a code attached in the correct order to a molecular backbone, thus making up a strand of DNA."[28] The sequence of these units in the DNA chain is a store of information which specifies the chemical structures that a cell can build. Thus, molecular genetics has revealed -- along with the fundamental knowledge as to how genes work and replicate -- a profound germinal concept: *that of molecular information transfer.*[29] Specific interactions between molecules in biological systems involve information not simply structure, chemical reactivity, or energetics. The unifying principles of molecular genetics are awesome in their presumed universality: (i) Genes are able to be reproduced, replicated. The replication of the linear array of genes in a chromosome can be explained by simple molecular rules. These same rules of replication apply to all genes, "whether in man, peas, fruit flies, or bacteria."[30] (ii) There is one genetic code which is used by all forms of life on earth. And yet the fundamental question persists: How can the linear sequence of bases in a segment of DNA provide the genetic information specific to a single gene?

Genetic Engineering. The discovery in 1973 that it is possible to recombine specific segments of DNA from different sources paved the way for recombinant DNA technology. Recombinant DNA technology and genetic engineering enable manipulation and altering of the genes in the laboratory and transfer of genes among different organisms. The gene pool of the earth, the life-determinant of the future, is now "the experimental subject for genetic engineering."[31]

The implications of these scientific advances for human disease are

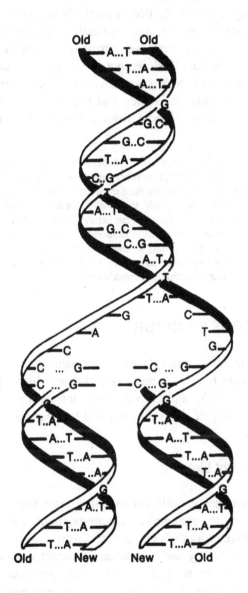

Figure 4.3. The DNA double helix (top) and the replication of DNA (bottom). When DNA replicates, the old strands unwind and serve as templates for new complementary strands. The new (daughter) molecules are exact copies of the old (parent), with each having one of the parent strands (from Ref. 27).

enormous. Over 2,000 hereditary disorders are known most of which are disabling and many fatal in early life. A particularly clear example of the cellular basis of disease is cancer, characterized by a persistent loss of some as yet unidentified normal genetic regulatory mechanism resulting in the over production of functionally defective cells. Equally profound are the dangerous implications of these same advances. Modern man's anguish regarding these developments is deep and has been expressed in many writings as in the following quote from a recent editorial in Science magazine:[32]

> "Nearly every other day it seems, a headline in our newspapers announces the discovery of yet another gene to some aspect of human biology, behavior, or disease.Perhaps most disturbing to our sense of being free individuals, capable to a large degree of shaping our character and minds, is the idea that our behavior, mental abilities, and mental health can be determined or destroyed by a segment of DNA. How much of our fate is in fact written in the DNA inside our cells? And how much freedom do we have to reach our full potential as human beings through our education and experiences?"

4.3. THE RICHNESS OF NATURE: LIGHT

One of the amazing characteristics of nature, observed Richard Feynman,[33] is the variety of interpretational schemes which is possible. The ways light is generated by and the ways light interacts with matter is an example of the validity of this statement and the immense diversity and richness of nature.

4.3.1. The Origins of Light

Light is electromagnetic radiation, radiant energy, and the term light is broadly used here to encompass electromagnetic radiation of all wavelengths. It moves with the same speed, c, everywhere in empty space independently of its energy and has the same characteristics everywhere. Indeed, light can never be put to rest. Once generated, it assumes instantaneously its speed. The speed of the electromagnetic waves in vacuum c is related to their wavelengths[34] λ and frequencies ν by $\nu = c/\lambda$. In any medium other than vacuum the speed of light is $\upsilon = c/n$, where n is the refractive index of the medium. Light quanta, the photons, have zero rest mass, energy $E = h\nu$, momentum $p = E/c$ and a particle-wave wavelength $\lambda = h/p$. Photons are absorbed by matter, scattered by matter, and interact with matter as particles do.

The wavelengths and the frequencies of light waves and the energies of light particles cover an enormous range (Figure 4.4). The various segments of the so-called electromagnetic spectrum refer to radiant energy normally associated with the way light in that particular region is generated. Light is called by different names depending on the specific process that led to its generation, or the way the light generating-system has been formed (excited).

Light can be produced by the acceleration of electrons or other charged particles. Commercial power lines radiate electromagnetic waves at the generator frequency (usually 60 Hz, but the frequency limit is practically zero). Vacuum tubes and other electronic devices generate electromagnetic radiation up to about 10^{11} Hz. This frequency range covers the applications to radio communication, radar, and commercial dielectric heating. Electromagnetic radiation higher than 10^{11} Hz is produced by natural processes as emission of photons from a multitude of excited species such as molecules, atoms, ions, and nuclei. The energy of the photon depends on the size of the emitting species in that the more confined the system -- say, the more tightly bound the electrons of the emitting species -- the larger the spacing between its quantum states and thus the larger the energies of the light quanta it emits. Independently of the photon energy, only systems which have excess energy and are thus transient can emit light.

- Light extending from about 25 μm to the red end of the visible spectrum at about 7000 Å is called infrared and is produced by radiative de-excitation of rotational and vibrational quantum states of molecules.

- Light extending from the red limit of human vision at about 7000 Å (1.8 eV) to about 4000 Å (3.1 eV) is called visible, and light in the wavelength range of about 4000 Å (3.1 eV) to about 2000 Å (6.2 eV) is called ultraviolet. It originates from radiative de-excitation of the loosely bound outermost electrons in molecules and some atoms.

- Light extending from about 2000 Å to about 100 Å is known as vacuum ultraviolet and is emitted by de-excitation of the more tightly bound electrons for which the energy spacings between the electronic quantum states is larger than for the outermost electrons. Vacuum ultraviolet light is invisible. It is strongly absorbed by matter, even by a very short path of air. Its study requires vacuum to free the equipment from the presence of air molecules which would otherwise absorb light of this wavelength.

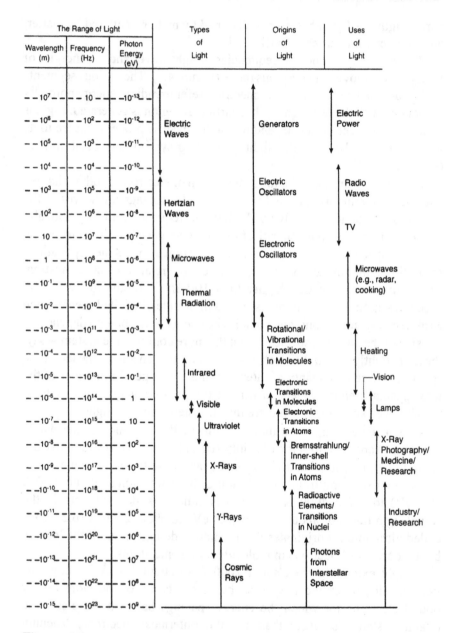

Figure 4.4. Range of radiant (photon) energy, wavelength, and frequency. The numbers for the wavelength and the frequency in the figure should be multiplied, respectively, by 1.2 and 2.4 (these multiplicative factors have been omitted for the convenience of display). Also shown in the figure are approximate regions of the various types, origins, and uses of light.

- Light emitted in atomic transitions of bound electrons between the inner (K-, L-, and M-) shells of atoms is known as characteristic X-rays.

- Light resulting from the acceleration of free electrons or other charged particles starting at about 100 eV and extending upward to the keV range is called Bremsstrahlung or continuous X-rays.

- Light originating from nuclear transitions or from radioactive nuclei is known as gamma rays and stretches in energy from about 10 keV to about 10 MeV.

- Light from interstellar space with energies reaching the GeV range is referred to as cosmic rays.

- Light generated by the process of electron-positron annihilation ($e^- + e^+ \rightarrow$ light) is known as annihilation radiation, light generated by the process of radiative attachment ($e^- + A \rightarrow A^- + h\nu$) is known as radiative attachment continuum, and light generated by the process of positive ion-negative ion recombination ($A^+ + A^- \rightarrow AB + h\nu$) and electron-positive ion recombination ($e^- + A^+ \rightarrow A + h\nu$) is known as recombination luminescence.

- Light generated in matter by the energy deposited in it by high-energy ionizing radiation (particulate or electromagnetic) is known as scintillations.

- Light generated when materials are excited by other photons is invariably called luminescence, fluorescence, phosphorescence, excimer fluorescence, resonance fluorescence and so on depending on the nature of the emitting species and the electronic state.[35]

- Light emitted by materials excited via chemical reactions, biological reactions, friction, heat, and an applied voltage is known, respectively, as chemiluminescence, bioluminescence, triboluminescence, thermoluminescence, and electroluminescence.

- Light emitted by all bodies by virtue of their temperature (that is, by all bodies not at absolute zero) is called blackbody radiation.

Modern science unraveled these and other light-generating processes and has recently found and developed unique ways of generating light with special properties. Two such new ways are the laser and the synchrotron light sources. The intensity, monochromaticity, and coherence of the former, and the broad and higher-energy spectral range of the latter opened up new opportunities for pure and applied science.

4.3.2. The Interactions of Light

Photons are emitted by every species. Photons are also absorbed by every species. They are, as we have discussed earlier in this chapter, the particles which mediate the electromagnetic interactions in nature. In principle, every process that generates light can be reversed and absorb light. Light can be absorbed by an atom, or a molecule, or a nucleus, or by bulk matter, or by any other species, and disappear. Low-energy photons (generally those with energies below the minimum energy required to ionize the species they interact with) are readily absorbed by the various substances in ways which depend on their absorption spectra, that is, the "allowed" transitions which the photons can induce. Since these transitions are uniquely characteristic of the atoms and the molecules themselves, the absorption spectra of atoms and molecules represent their unique signatures, their fingerprints. Each species is a unique filter of the light that impinges on it.

When photons are absorbed by molecules, rotations may be excited, vibrations may be excited, electronic transitions may be excited, one or more electrons may be excited or ejected, the molecule may be dissociated into neutral fragments and/or charged fragments, and so on. Higher energy photons have additional ways of interacting with matter. For instance, photons in the energy range of 1 keV to about 50 MeV can interact via the photoelectric effect process where the photon is absorbed and an electron is ejected from the absorbing species, or via the Compton scattering process where the photon is scattered by a free electron of the target and an electron is ejected, or via the pair-production process when their energy is at least equal to 1.02 MeV $(= 2\, m_o c^2)$. In this last process the photon is completely absorbed by matter and its energy converted into the rest-mass-energy $(2\, m_0 c^2)$ of a positron-electron pair and their kinetic energy. The pair-production process is thus the reverse of the positron-electron annihilation. In the former process light disappears and in its place matter is generated. In the latter process matter is transformed into radiant energy.

- Light can be absorbed by atoms or molecules to produce an electron-positive ion pair $(h\nu + A \rightarrow A^+ + e^-)$, or by molecules to produce a negative ion-positive ion pair $(h\nu + AB \rightarrow A^- + B^+)$, or by a negative ion to photodetach the extra electron it carries $(A^- + h\nu \rightarrow A + e^-)$, just as light is produced by the reverse processes of electron-positive ion recombination, negative ion-positive ion recombination, and radiative attachment discussed above.

- Light can be absorbed by bulk matter collectively and by macromolecular structures which may rupture as a result of its absorption.

- Light interacts with light, with electric fields, and with magnetic fields.

- Light is absorbed by chlorophyll (the major light-absorbing pigment in plants) and is converted into chemical energy.

4.3.3. What Would We Do Without Light?

We would do very little, if anything. Light supports life through photochemical phenomena. Life on earth depends on photosynthesis, a series of processes in which electromagnetic energy is converted to chemical free energy which can be used for biosynthesis. Through photosynthesis plants convert solar energy into chemical energy and add an estimated[36] 10 billion tons of carbon into the biosphere annually as carbohydrate.

Light warms the earth and the life on it. Infrared radiation is mainly responsible for the heating effect of the sun. The molecules of our skin "resonate" at infrared frequencies and are thus preferentially absorbing the sun's infrared radiation warming us up.

Light images the world for us directly or indirectly through the instruments of science. It sustains life in a multitude of ways both visible and invisible. We can see objects in the world around us because light bounces off them and enters our eyes. The rays of light create an image on the retina and stimulate nerve cells sending electrical impulses to the brain. The human eye responds to visible light with extreme sensitivity. The eye uses very effectively the photons caught by the visual pigments of the rodes and cones. A rod can be excited by a single photon.[37]

Light is at the basis of modern man's communication systems, scientific systems, and technological systems.

And while we may understand the entirety of the electromagnetic spectrum intellectually for wavelengths ranging from 10^{-16} m to 10^8 m, it still remains a fact of human existence that of all these wavelengths the most wondrous and the most beautiful are those in a tiny range from 7000 Å to 4000 Å -- the visible region -- to which the traditional term light normally refers. It is in the visible light that the beauty of light is normally visible to man.

Light! Beautiful and wonderful light! It is no wonder that in many religions light is used in real or allegorical ways to best describe the presence or essence of God.[38]

4.4. THE LAWS OF NATURE

The most distinct characteristic of science is its established laws, the descriptions of regulations in the observed behavior of natural phenomena. From them derives science's predictive power, the ability to make statements about the regions of the physical world that have not been seen based on the established laws of science. Let us look at just a few characteristics of the physical law.

4.4.1. The Generality of the Physical Law

The physical laws established by science have universal validity in the sense that they hold in every part of the universe so far investigated. They are the same no matter when and where they are established.[39] Matter, energy, and light behave the same way on earth as they do in outer space. Matter is composed of the same particles and the same elements everywhere. Light moves with the same speed and has the same characteristics everywhere. Photons, electrons, protons, neutrons, the elements and their nuclei are the same everywhere. In the general applicability of the physical law implicitly lies the notion of the uniformity of nature. And yet one may ask: are there periods in the history of the cosmos when, or regions of the cosmos where, this notion of the immutability of the physical law is invalid?

The generality of the established physical law is a powerful instrument in discovery. We shall illustrate this by two examples, the established symmetry of the physical law and the established laws of conservation of fundamental physical quantities such as energy, momentum, and charge.

4.4.2. The Symmetry in the Physical Law

According to Hermann Weyl a thing is symmetrical "if there is something that you can do to it so that after you finished doing it it looks the same as it did before."[41] It is in that sense that the laws of physics are said to be symmetrical, "that there are things we can do to the physical laws, or to our way of representing the physical laws, which make no difference, and leave everything unchanged in its effects."[40] The simplest example of this kind of symmetry is displacement in space (that is, the equivalence of directions and of different points of space) and a similar one involving displacement in time.

Symmetry laws are frequently used to check the validity of theoretical propositions or to device new constructs. For instance, the construct of wave-particle duality was proposed on symmetry arguments: if a wave exhibits particle-like properties, a particle should also exhibit wave-like properties. Similarly, the realization that for ordinary matter to be stable there must exist a mirror image of it led to the prediction of antimatter, the complementary of matter. Symmetry is the primary guide into the structure of elementary particles even though high-energy physics found new features of the natural laws which are exceptions, such as the violation of parity or "mirror symmetry." The reversibility of many microscopic natural processes is another example of the usefulness of symmetry in discovery. What goes in one direction must go the reverse way unless something else interferes. The symmetry in the reversibility of microscopic processes enables the scientist to obtain knowledge on the forward reaction from knowledge on the backward reaction and vice versa. For example, from knowledge on the absorption of light by one electronic state of an atom raising it to another we can obtain information on the emission of light from the second (higher) electronic state back to the original state, from knowledge on the photoionization of an atom we can obtain information about the reverse process of recombination, from knowledge on the photodetachment process we can obtain information about the reverse process of radiative attachment, and so on. More generally, the law of gravitation, the law of electromagnetism, and the laws of nuclear reactions are time-reversible, although the phenomena of the world are evidently irreversible. And lets we forget, we are reminded by Wigner[39] that symmetry applies to the laws of nature, not to the events themselves.

4.4.3. The Conservation Laws

Science has found that certain quantities are conserved in nature. Foremost among these is energy (mass/energy), momentum (linear and angular), and charge.[41]Energy has all kinds of forms and is equivalent to mass. It is conserved and so is the total momentum (linear and angular) of a system and the total electric charge in the world.

New phenomena have to satisfy the laws of conservation of physics. The universal validity of the conservation laws has been masterfully used as an instrument in discovery. Thus, the law of conservation of angular momentum was instrumental in the construction of Bohr's atomic model, and the laws of conservation of energy and momentum are the basis of the discovery of many new elementary particles. An additional early example of this which we referred to in the previous chapter is the discovery of the neutrino in the disintegration of a neutron into a proton and an electron. The fact that the energy in such a reaction did not check out was the basis for the prediction of the new particle, the neutrino, later found to exist.

The conservation principles are obeyed by all physical laws. They are distinctly universal. Along with the other general physical laws established by science, they extend the penetration of science over incredible ranges of times and distances varying by 10^{40} orders of magnitude (see Figures 1.2 and Figure 1.3 in Chapter 1).

4.5. FUNDAMENTAL GENERALIZATIONS AND QUESTIONS

Let us close this chapter by recounting a few generalizations in man's knowledge of the physical universe that have been made possible by science and which we have eluded to in the preceding chapters.

All matter is composed of the same particles and elements everywhere. The matter of which the stars are made and the matter of which the earth is made is the same. The same kinds of atoms appear to be in living and non-living entities; they are just differently arranged. The atoms themselves are made from the same general constitution everywhere. Electrons, protons, neutrons, the nuclei of the elements are the same everywhere. The nucleon is the basis of all matter. Thus, science has created a framework for a unified description of the natural world on a microcosmic and on a macrocosmic level. This framework allows us to see fundamental connections between the properties of nuclei, atoms, molecules, living cells, and stars. It tells us in terms of the constants of nature why matter in its different forms exhibits the qualities we observe.

Science speaks of relative time and its connection with space. It speaks of universal principles such as gravitation and conservation of energy and charge. It speaks of electric and magnetic fields, of forces, of fundamental units of matter, and of the living cells as parts of physical reality. It speaks of energy which is not only conserved, but which is constantly being dissipated, accumulated, and transformed. It speaks of the universality of the genetic code; the cellular structure, reproduction, and evolution of living forms; the common character of the proteins in different species and the fundamental similarity of energy of metabolism in all species.

Yet this knowledge is neither complete nor static. It is presently not sufficient either as a body of knowledge or as a process of inductive inference to explain the complexity of macroscopic and living matter. There exist questions beyond the power of the present physical law raised in one form or another by scientists and non-scientists alike. Questions such as: What is the deeper meaning of the physical laws? What is the significance of the particles and subparticles of which matter is composed? What is the origin of matter? How do growing organisms develop their complex structures?

And so it seems that the penetrator of the physical universe leaves unanswered essential questions. It speaks of a primordial explosion, of primordial fires, of a primordial biological material, and of an absolute and unsurpassable speed, and continues unabated its search for better answers knowing that each answer will come with yet more difficult questions.

4.6. REFERENCES AND NOTES

1. T. T. Tsong, Physics Today, May 1993, p. 24.
2. G. Binning, H. Rohrer, Ch. Gerber, and E. Weibel, Phys. Rev. Lett. **50**, 120 (1983).
3. G. Binning and H. Rohrer, Scientific American **253**, August 1985, p. 50.
4. C. F. Quate, Physics Today, August 1986, p. 26.
5. H. K. Wickramasinghe, Scientific American, October 1989, p. 98.
6. C. Bustamante and D. Keller, Physics Today, December 1995, p. 32.
7. *To the Heart of Matter-The Superconducting Supercollider.* University Research Association, Washington, D. C., 1990.
8. M. Riordan, Science **256**, 29 May 1992, p. 1287.
9. J. D. Barrow and J. Silk, *The Left Hand of Creation,* Oxford University Press, New York, 1983, p. 219.
10. Particles and antiparticles annihilate one another whenever they collide and energy is created in their place. For instance, when an electron and its antiparticle, the positron, meet they disappear and a pair of gamma rays (annihilation quanta) is created. The reaction is an example of the transformation of matter into energy.
11. P. Renteln, American Scientist **79**, November/December 1991, p. 508.
12. F. Close, *The Cosmic Onion,* American Institute of Physics, New York, 1988.
13. The W and Z bosons were experimentally observed in 1983 at CERN. Their existence had   been predicted by the electroweak theory according to which the electromagnetic and weak forces are different manifestations of a single unifying interaction.
14. The National Research Council, *Science and Technology,* W. H. Freeman and Company, San Francisco, 1979, p. 151.
15. D. J. Helfand, Physics Today, November 1995, p. 58.
16. J. D. Burrow, in *Physics and our View of the World,* J. Hilgevoord (Ed.), Cambridge University Press, New York, 1994, p. 49.
17. Ref. 9, p. 31.
18. G. Lake, Science **224**, 18 May 1984, p. 675.

19. The National Research Council, *Science and Technology*, W. H. Freeman and Company, San Francisco, 1979, p. 157.

20. Ref. 9, p. 43.

21. The National Research Council, *Science and Technology*, W. H. Freeman and Company, San Francisco, 1979, p. 154.

22. S. W. Hawking and G. F. R. Ellis, *The Large Scale Structure of Space-Time*, Cambridge University Press, Cambridge, 1973.

23. S. L. Shapiro and S. A. Teukolsky, American Scientist 79, July/August, 1991, p. 330.

24. Ref. 9, p. 89.

25. Ref. 9, p. xiii.

26. T. H. Geballe, Physics Today, November 1981, p. 132.

27. Office of Technology Assessment, *Impacts of Applied Genetics*, Washington, D. C., 1981, p. 39.

28. D. Braben, *To Be a Scientist*, Oxford University Press, Oxford, 1994, p. 32.

29. The National Research Council, *Science and Technology*, W. H. Freeman and Company, San Francisco, 1979, pp. 60 and 61.

30. D. S. Hogness, in *The Greatest Adventure*, E. H. Kone and H. J. Jordan (Eds.), The Rockefeller University Press, New York, 1974, pp. 141-147.

31. L. F. Cavalieri, The Bulletin of the Atomic Scientists, December 1982, p. 72.

32. T. N. Wiesel, Science **264**, 17 June 1994, p. 1647.

33. R. P. Feynman, *The Character of Physical Law*, The M.I.T. Press, Cambridge, MA, 1967.

34. The wavelength of an electromagnetic wave is the distance measured along the line of propagation between the crests or any two points which are in phase on adjacent waves. The number of waves per centimeter is called the wavenumber. Frequency is the number of waves that pass a particular point per unit time or the number of cycles per unit time, usually per second.

35. L. G. Christophorou, *Atomic and Molecular Radiation Physics*, Wiley-Interscience, New York, 1971, Chapter 3.

36. G. R. Fleming and R. van Grondelle, Physics Today, February 1994, p. 48.

37. W. A. H. Rushton, in *Light and Life*, W. D. McElroy and B. Glass, The Johns Hopkins Press, Baltimore, 1961, p. 707.

38. V. Lossky, *In the Image and Likeness of God*, St. Vladimir's Seminary Press, Crestwood, New York 10707, 1985.

39. E. P. Wigner, Physics Today, March 1964, p. 34.

40. Ref. 33, pp. 84 and 85.

41. Another such law is the law of conservation of baryons. If the neutron and the proton is each counted as one unit, or baryon, then the number of baryons in nuclear reactions is conserved.

CHAPTER 5

Distinct Characteristics and Principles of Science

The emergence of big science has created many and difficult problems for science, the institutions of science, and the relation of both to society. Bigness and economic, political, and public pressures have caused a loosening in the coherence of modern science and a loss of science's traditional unity of purpose. This has diminished the ability of science to shield itself from outside encroachments. Fortunately, there are certain distinct characteristics and principles of science which are embedded in its tradition and method that still qualify many of its functions. Even today, these characteristics and these principles are valid. They need to be recognized and adhered to if science is to serve well and in a balanced way man and his societies.

5.1. DISTINCT CHARACTERISTICS OF SCIENCE

Science has many and distinct characteristics and here we elaborate on just a few.

Science is an Adventure. Other desires die in their gratification, it has been said, but not knowledge. The thirst for knowledge is never quenched. Science engages man in a perpetual adventure which emphasizes and challenges his intellect. It satisfies his natural curiosity to know, urges him to discover the natural order, and dares him to unify the seemingly unlike. This was indeed the feeling of the great synthesizers (unifiers) of science, such as Faraday when he linked electricity and magnetism, Maxwell when he linked electricity and magnetism to light, and Einstein when he linked mass with energy and time with space. The unification of the fundamental forces of nature and the unification of the physical and the biological phenomena are perhaps two of the most distinct elements of this adventure today. The scientist consciously accepts the challenge of the unknown, but knowable world. Unraveling the secrets of nature keeps him at the edge of mystery. Bewildered, he ventures into and confronts the unknown, forming an intimate bond between himself and nature; an interwoven intimacy from which flaws awe, responsibility, and supreme ecology. A scientist's embrace of nature becomes an extension of himself. Modern science, wrote Whitehead,"has imposed on humanity the necessity for wandering. Its progressive thought and its progressive technology make the transition through time a true migration into uncharted seas of adventure."[1]

Science is Beautiful. Neither the prophets of religion in their attempts to uncover the glory of God in the beauty of His creation, nor the artists, nor the poets, and nor the philosophers of cultures past in their attempts to paint or express the immensity of natural beauty were ever able to sketch its richness as has been done by science. There is amazing beauty in what is perceived by the senses, but there is also unimaginable beauty the normal senses cannot sense. Science increasingly presents objects, concepts, and worlds that are beyond the response and receptivity of the ordinary senses. The beauty in what the eye can see is breathless and yet tiny in comparison with the boundless beauty beyond the eye's limited vision; infinite beauty lies beyond the visible light. Indeed, most of today's science is beyond normal "seeing." Through science we learn to know without seeing and what we "see" is beautiful however indirect is our contact with it.

Whether contemporary science provides man a picture of nature or a picture of his "relation to nature"[2] by interpreting images of reality, man's indirect contact with the physical world presented by science is real just the same. And there is beauty in the richness of nature's phenomena and processes. How enriching it would have been for the non-scientist to "see," for instance, the multitude of ways in which light can be generated, interact with the microscopic, macroscopic, and living matter and organisms, and induce new phenomena and conditions for life (Section 4.3). The beauty of regularity and intricate order stretching from the macrocosmos down to the known forms of bulk matter, to the fundamental elements of the microcosmos -- the molecule, the atom, the nucleus, the nucleon, the quark. The beauty in symmetry and antisymmetry, in order and disorder, in conformations and transformations, in stable and transient structures, in the simple and the complex, in the small and the big, and in the living. Amazingly elegant beauty. "Was it a God that wrote these signs?" asked Ludwig Boltzmann overwhelmed by the elegance of Maxwell's equations.[3] Regrettably, it is difficult for the non-scientist to partake fully of this beauty. Every person, nonetheless, can marvel at the majestic beauty of the "laws"[4] of nature and their predicting power and recognize that even more magnificent than these laws is the miracle of man's intellectual power which enables him to discover and comprehend them. Indeed, the beauty of the unraveled cosmos is surpassed by the miracle of man's ability to unravel it.

Science is not Retrograde. Science is by far the largest producer of new knowledge today. It is thus the largest agent of expansion of man's world because "every time we acquire knowledge we enlarge the world of man."[5] "What is once acquired of real knowledge can never be lost," wrote Thomas Jefferson.[6] When a fundamental law of nature is discovered, it is discovered for all time. It can be modified, expanded, improved, or exhumed, but it will never be discovered in kind again. Every scientific discovery is unique: it happens only once in the form it originally did and it affects the next level of discovery only in the way it did. Science grows like a tree, it has been said,[7] ring by ring. Yet, paradoxically, there is *no completality* in science. Science is open-ended. Scientific knowledge being not absolute, is always incomplete and is always emerging. Discovery breeds discovery and novelty supersedes novelty. In a chain-like process, ideas about nature are reconstructed and yet remain incomplete. They are valid within shrinking limits of applicability. For each of the realms of

Newtonian mechanics, statistical thermodynamics, special theory of relativity, Maxwell's electrodynamics, and modern quantum mechanics, wrote Heisenberg, "there is a precisely formulated system of concepts and axioms, whose propositions are strictly valid within the particular realm of experience they describe."[8] Whether in genetics or in quantum mechanics or in cosmology, what was held firm 80 to 100 years ago is not held so now and what is now accepted will, most likely, not be so accepted 100 years hence. Each step is nonetheless a frontier at a given time, a necessary link to a clearer, yet incomplete, comprehension. Indeed, some theories have a distinctly provisional character. One such example is the first theory of the atom by Niels Bohr. It was provincial, meant to be used as a scaffold and in time to yield to a more refine theoretical structure. It was indeed superseded by modern quantum theory. It is a milestone just the same in the historical evolution of man's ideas about the structure of the atom.

Science is Pragmatic. It uses common sense and deals mostly with what works or has a chance to succeed and is held important both within and outside science. This diverts science from intractable questions (which may, at times, be the most important) and embraces the world of *relevance* and *usefulness*. Science is pragmatic in yet another way: if an experiment or an approach is proved wrong, or if the results of an experiment are different from what has been anticipated, science is content to accept the verdict, pleased that this has been proved to be so. Falsification of scientific facts is a mortal sin in science and although in rare occasions it does surface it cannot escape detection for too long because *verifiability* is a distinct characteristic of science. The history of science reveals that no fraud or honest mistake will stay undetected for too long. Scientific criticism is inherent in the scientific process itself because science forces every scientist and every science worker (Chapter 6) to turn to nature as the arbitrator of truth. *Science is a self-correcting system.*

Science is pragmatic in three other unique ways. First, *science is independent of the individual scientist.* What Michelangelo or Beethoven created would not be created if they did not do it, but what Newton, Einstein, or Rutherford discovered would have been discovered by someone else sooner or later. "If Einstein had not discovered relativity theory," wrote Heisenberg,[9] "it would have been discovered sooner or later by someone else, perhaps Poincaré or Lorentz. If Hahn had not discovered uranium fission, perhaps Fermi or Joliot would have hit upon it a few years later. I don't think we detract from the great achievement of the individual if we express these views." Indeed, we do not. The scientist has more reasons to be humble than does the artist or the composer! Second, *science knows the nature of the knowledge it provides.* "Order must be discovered and, in a deep sense, it must be created," noted Bronowski.[10] "For the purpose of acquiring knowledge, it is always necessary to find simple and general relations and to select such objects as display relations of this sort. Consequently, it has been found necessary for many scientific purposes to replace the perceptual objects of common sense by *scientific objects* which are conceived as composing the perceptual objects. These scientific objects are Particles, Molecules, Atoms, Electrons, ... The great advantage that is gained by the use of scientific objects is that their relations to one another and to events are simple and general," wrote Ritchie.[11] "In science the object of research is no longer nature in itself but rather nature exposed to man's questioning. .... The mathematical formulas indeed no longer portray nature, but rather our knowledge of nature," contented Heisenberg.[2] Modern science is "an artifact," it is "man-made" reasoned Rabi.[12] Third, *science knows its boundaries.* It describes the structure of matter, it does not discover values or norms. "It has nothing to say about 'the meaning of the world,' about the 'meaning of existence,' or about the 'aims of society.' This is where the boundaries of science lie. They ought to be respected."[13]

Science is Unpredictable. Who would have predicted the existence of the atomic nucleus when Rutherford began his experiments with alpha-particles, and who would blame him from failing to predict that out of his discovery of nuclear transformations would emerge nuclear power? Science is as variable and unpredictable as is the human spirit. The physical and the chemical reactions are not obligated to conform to the scientist's wishes. And they do not. X-rays, penicillin, recombinant DNA altered the face of medicine and they all came from efforts unrelated to medicine. Yet, one of the most persistent and distinct characteristics of scientific knowledge is *regularity* and *predictability.* Science is concerned with what is *common* among events. It seeks regularity, relationships, and laws linking phenomena and observations, and attempts to explain these from basic principles. To accomplish this, science establishes methods and techniques to interrogate nature, to provoke the universe to reveal itself, and to weed out the proliferation of ideas. Through critical, methodical, and persistent testing of alternatives, science resolves ambiguities and *establishes the facts.* Without established facts little progress can be made. The regularity of scientific knowledge is a great economy of thought. Through it is achieved understanding which is a prerequisite of prediction. The power and the capacity to predict is derived from established regularities and is a distinct characteristic of scientific knowledge. Belief that the course of nature is regular, maintained Russell,[14] gives a sense of security. Predictability is closely related to simplicity, although complexity is generally required to explain detail.

Science is Utilitarian. Inherent in science is the vitality and indeed the tendency to multiply itself in whatever form, and in whatever form it multiplies, its main body of discovery ultimately achieves practical utility even in those cases where the discoveries were made initially solely in the interest of *pure* science. It suffices to illustrate this by two examples. First, when Ernst Rutherford was asked in 1933 if he foresaw any applications from his discovery of nuclear transformations, he characterized[15] such a prospect as "moonshine." Barely ten years later Enrico Fermi proved the master completely wrong when he had demonstrated the first nuclear chain reaction in Chicago. Very few "pure science" discoveries to date impact human life more than those which led to the atomic bomb and the nuclear reactor. The second example is that regarding the discovery of the laser. When A. L. Schawlow, the co-inventor of the laser principle, was asked a similar question about his laser invention, he described it as "an invention in search of an application."[16] Yet, the invention of the laser has developed into one of the most important technological discoveries of the twentieth century. The laser has become one of the most powerful tools the modern scientist has at his disposal to study nature and the modern engineer uses lasers to create new technologies whether in materials processing, communications, optical data processing, or medical applications.

Many other examples of science-based technologies can be cited (Chapter 7) that have changed and will continue to change the conditions of modern life. But lest we forget, the television set, the transistor radio, the computer, the miracles in communication and transportation systems, the nylons and the myriad of synthetic materials, the marvels of biomedicine, and the effectiveness of modern warfare can remind us of the utilitarian quality of science. Technological changes since the late 19th century are all rooted in scientific discovery. Discovery upon discovery led to new breakthroughs, new inventions, and new technologies. The marriage of modern science and technology has been, is, and will remain strenuous, but it will become more intimate in the future as science will continue to affect man in new ways, directly, but also subtly and obliquely by facilitating the "filters" through which man views new aspects of reality.

Science is Conservative and Revolutionary. Science, wrote Weisskopf, "is both traditional and revolutionary at the same time. Newton's mechanics and the electrodynamics of Faraday and Maxwell are still valid and alive. Current calculations of satellite orbits and of radio waves are still based on them. Revolutionary concepts, such as that of relativity and quantum theory did not invalidate the earlier ideas; they establish unexpected limitations to the old ideas, which remain valid within these limitations. On the other hand, there is a strong trend in science toward the new and the different. .... But scientific revolutions are extensions rather than replacements."[15] Others[17] maintain just the opposite. That scientific revolutions are extensions rather than replacements can be seen from the examples we gave in Chapter 3 on the embryology of science and the evolution of scientific knowledge. Nonetheless, time and again, new discoveries take place that shake the accepted view. Although unanimity has no status as a means of proof in science, at all times there is a predominantly accepted scientific view, a paradigm,[17] within which the scientist and the science worker conduct their vocation. Facts or theories contradicting this accepted view are treated with caution and are often met with resistance. While this accounts for the bitter opposition encountered by many new scientific truths prior to their ultimate acceptance, it demonstrates the complementary character of science of being simultaneously revolutionary and conservative. Each change necessarily encompasses previous knowledge.

Science is not Skepticism. In science it is honorable to doubt, but it is not the practice of science to look for things to doubt. "Perpetual doubting and a perpetual questioning of the truth of what we have learned is not the temper of science."[18] Recognition of the significance of facts that were previously overlooked and disregard of facts that were previously considered important has led to major new discoveries, but it is not by deliberate skepticism that discoveries are made. To quote Oppenheimer once again, "it was not by a deliberate attempt of skepticism that physicists were led to doubt the absolute nature of simultaneity, or to recognize that the ideas of strict causality embodied in classical physics could not be applied in the domain of the atomic phenomena."[19]

Science is not Hierarchical. Paradoxically, in science there is neither democracy nor monopoly. Not all opinions carry the same weight, and yet axiomatization and dogmatism mean sterility and are rejected. The opinion of the young can outweigh the opinion of the elder, and the junior's discovery can overshadow the senior's accomplishment. There is no substitute for either the boldness of the young or the intuition of the more mature scientist. Science rests, in the long run, on the consensus of the informed opinion of the scientists, not on the authority of any one individual, no matter how outstanding. And, yes, there is no place for political correctness in the facts or the functions of science.

Science is Cooperative. The spectacular recent scientific and technological achievements are the net result of long and continuous efforts made possible once a critical amount of relevant knowledge in science and technology has accumulated. Classic examples remain the space program, the moon landing missions, and the Manhattan Project in the U.S.A. that built the first atomic bomb. Science's recent past distinctly demonstrates the cooperative nature of the scientific endeavor and the collaborative character of many recent scientific accomplishments. This is so not just for the projects of "Big Science" such as those in particle physics (e.g., European Commission of Nuclear Research, CERN), molecular biology (e.g., the Genome Project, the international effort to map the human genome), the international efforts relating to the environment (e.g., Global Atmospheric Research Program), the research efforts in Antarctica and so on, but also for the numerous and diverse projects of "Little Science," such as those in material science and the physical, chemical, and biological effects of ionizing radiation. However, it can be argued, that the emergence of team research efforts and the birth of hybrid disciplines (for instance, chemical physics, nuclear medicine, bioengineering), as well as the systematic inquiry in pursuit of new knowledge are rather recent characteristics of science.

Science is Universal. Science is universal in its method, in its validity and validation, in its language, in its mythopoetic functions, and in its effects on and compression of all human functions. The same topics are discussed and researched everywhere. Science has built upon the discoveries of all, and is, thus, the work of many people. Men and women from all nations have learned to do science together and in this regard science stands as an example of what men and women can accomplish by working together. "The structure of modern science is the greatest collective achievement of the human mind. To build it has required patience and vision, imagination and discipline, cooperation and honesty. To study it and to add to it will foster these qualities."[20] These qualities are necessary because science *is a liberating force for mankind.* From the time the second law of thermodynamics allowed the steam engine to replace slavery, to the development of today's energy sources, materials, and medicine, science has liberated man from the bondage of need, want, fear, and prejudice. Indeed, if deep in the essence of civilization lies the emancipation of humanity, society cannot be truly civilized without science. Science is a liberating force in society not only because of what it does for its material needs, but also because of what it does for the mind and the spirit of man and for his political freedom. It is, for example, through the means science enabled technology to develop that information penetrated and broke up the "iron curtain" of totalitarian states, liberated oppressed peoples, ended their isolation, and exposed human suffering.

It is as well necessary to recognize that while science is a facilitator of freedom, it itself requires freedom to optimize its functions. Science's past teaches that it flourished under a variety of social, economic, and political systems. However, due to its revolutionary character, science suffers in static and rigid environments such as those of the Middle Ages and the recent totalitarian regimes.[21] Science can be forced to comply and to restrict itself, but it cannot be blocked. Any time and for whatever reason though science is forced to discontinue its functions, it suffers a serious setback. Science cannot be stopped and be expected to take off again with the same vigor at a later time. The universality, international character, and power of modern science, afforded it considerable immunity from political authority and religious dogma. Paradoxically, even under totalitarian regimes, science flourished when it served the purposes of the ruling class, and was nurtured for that purpose. The strength of Soviet physics during the Soviet A-bomb and H-bomb development,[22] and the Soviet space program are two such examples.

Science is Society's Heritage of Common Knowledge. Science is universal. As such, it transcends national boundaries and generations. It belongs to all humanity at all times. Generation upon generation builds on all of mankind's prior accomplishments. And although most of the scientific research today is still done -- and most of the profit from the latest advances in science is still concentrated -- in the developed nations of North America, Europe, and Japan, scientific knowledge is available to everyone; modern communication systems have made this a reality. The heritage of common knowledge provided by science becomes a uniting force of humanity and a heritage of universal hope in spite the lingering pressures of nationalism.

5.2. GENERAL PRINCIPLES OF SCIENCE

As a human activity science has its tradition. Persistent qualities -- the principles of science -- characterize this tradition. These principles provide guidance for the functioning of science and link the scientific method to the rest of human thought and conduct. They qualify the place of science in the world of values and facts. There are values at the base of science arrived at non-scientifically, it has been pointed out by many. Values which guide the execution of scientific research, qualify the conduct of the scientist, define the scientific norms, and grand science a degree of continuity, cohesion, and unity. On the responsible freedom of the scientist depends the integrity of science, and on the integrity of science ultimately depends "the welfare and safety of mankind."[23]

We identify below nine principles of science: The Principle of Explanation, The Principle of Scientific Parsimony, The Principle of the Universality of the Physical Law, The Principle of Relatedness, The Principle of Embeddedness, The Coordinating Principle, The Principle of Correspondence, The Principle of Complementarity, and The Principle of Continuity.

The Principle of Explanation. This principle is rooted in the essence of rationality. It is based on the premise that nature is rational and the world is intelligible. The universe has to be rational to be intelligible and it has to be intelligible to be understood. The principle of explanation embodies the belief in a deep structure of reality to be revealed by scientific inquiry. A belief that the physical world is accessible to science because there are causal laws to be discovered. There is thus an "anthropic character"[24] in the scientific conceptualization and rationalization of nature.

The Principle of Scientific Parsimony. Thales of Miletus is credited for having founded the principle of parsimony: a maximum of phenomena should be explained by a minimum of hypotheses. Out of this principle emerges a basic criterion of truth: the ultimate simplicity of the physical law. Newton's equation of motion ($F = ma$) and Einstein's mass-energy equation ($E = m\,c^2$) are two classic such examples to which might be added the conservation laws of physics. Clearly, not all physical laws are that simple, but they are majestically elegant when they are.

In science, the principle of parsimony embodies in a general way the notion that truth is whole and that, therefore, the aim of science is to unify man's picture of the physical world and enable an explanation of the world with the fewest *ad hoc postulates*. Thus, commonly practiced in science is the *assessment of plausibility*.[25] In science the validity of the laws of nature has been established without *omniscience*[26] (i.e., without knowledge of each and every individual relevant event). Therefore, the purpose of science, as Peter Medawar said, is to free us from the tyranny of the particular. Simplicity and information compression are necessary for economy of thought, better understanding, wider and more predictive capacity. Beyond utility -- and in spite of the fact that the road to nature's simplicity normally passes through a great deal of complexity -- there is a deep mystery in the beauty of simplicity. "What is it," asked Richard Feynman, "about nature that lets this happen, that it is possible to guess from one part what the rest is going to do? ... I do not know how to answer.... I think *it is because nature has a simplicity and therefore a great beauty.*"[27,28]

The Principle of the Universality of the Physical Law. An aspect of the previous principle deals with the number of events a scientist has to study or consider to generalize a phenomenon or a relation. This principle states that the established laws of nature are universal. They extend in time and in space and are independent of the time interval between the detection of a phenomenon and its initial cause. The laws of nature established now on earth, have been, are, and will always remain applicable anywhere in the universe under the conditions the laws have been established. This is held applicable regardless of the fact that man's knowledge of the macrocosmos relies on signals the scientist detects on earth that had originated from stars (under virtually unknown conditions) billion of years earlier. Even regarding man's knowledge of the microcosmos the scientist deals with events which are separated in time by enormous time intervals. Large time intervals, for instance, separate cause and effect in the physical interaction of ionizing radiation with living organisms (e.g., man) and the ensuing biological/health effect (e.g., cancer). The energy deposition in the human body by the radiation is fast (it is completed within less than a fraction of a picosecond) and the effects of this energy deposition (e.g., cancer) are slow (they often appear years later) and yet the effect is traceable to the cause.

Whether in cosmology, in physics, in biology, or in anthropology, the scientist is asked to understand the old and the distant from the restricted and the now. The statement that the laws of nature are applicable everywhere in the cosmos is an axiomatic proposition that seems consistent with the observed phenomena. In spite of this and in spite of the temporal nature of the laws of science, the universality of the physical law is another distinct characteristic of nature's intrinsic simplicity and beauty. One needs point to the conservation principles of science. They are inviolable and standards against which newness is judged (see Chapter 4). They have been only rarely modified to encompass new conditions. They constitute frames of reference against which the behavior of nature is assessed. What violates them is unacceptable as suggested, or is profound.

The Principle of Relatedness. A necessary condition of all life is interdependence. Everything relates to everything else, to the whole. Science seeks not only truth, but also *relatedness*.[29] The various parts of science must illuminate and cohesively bind to their neighbors. Each basic scientific activity must strengthen the unity of science and the soundness of human knowledge. The laws of nature apply across all of science. They, thus, transcend the individual scientific disciplines and should be used to strengthen their activities, especially their esoteric and remote (isolated) parts. Recently, with the extreme diversification, proliferation, esoteric, and often imprisoned (via secrecy, proprietary restrictions, or shear competitiveness for priority) forms of scientific inquiry, this principle assumes special significance.

The Principle of Embeddedness. Everything relates to everything else, says the previous principle. Nothing exists in isolation. Nothing can even be defined in isolation. Everything assumes essence in its interactions with the something else. Man, thus, is never a privileged observer outside of the system he lives in and is a part of. He acts implicitly. Man's actions are embedded within those of his fellowmen. Out of this implicit embeddedness stems the need for respect and mutual accommodation and emerges the concept of brotherhood, which itself is based on mutually coupled relationships, interests, and values. We include values because what in practice holds for -- and determines -- our actions, also holds for -- and determines -- our values: *our values' value is implicit.* Values owe their value to the existence of other values. We cannot explain or value values without assuming or accepting (e. g., transcendent ethical religious values) some value. Each of the concepts of justice, freedom, dignity, love, and so on, is qualified by the rest and often with reference to an external standard. For instance, the extent to which we value justice depends on the degree to which we value man's life, civil and human rights. The extent to which we respect individual rights depends on the degree to which we value individual freedom. The extent to which we respect our fellow man qualifies our tolerance for his traditional ways of life. The love we have for our country qualifies our respect for another's love of his own country. Our love for God mirrors our love for man. This embeddedness of values allows for their mutual interaction and feedback, their mutual accommodation and indebtedness. It demonstrates the existence of an underlined unity amongst the values of man. This realization is an element of hope for it demonstrates that on the fundamental level man can converge on a universally acceptable value judgement system.

In science, the principle of embeddedness expresses the merit of relatedness of one branch of science to another. Out of this implicit coupling of the parts of science emerges the underlined unity among its seemingly chaotic functions. The mutual embeddedness of the parts of science allows for their integration, feedback, and accommodation. Thus, the embeddedness of biology and physics can be exemplified by the development of the Watson-Crick model of DNA which would have been impossible without major advances in X-ray diffraction. Many advances in molecular biology would not have been possible without radiation tracers. Similarly, the embeddedness of chemistry and physics can be exemplified by the many physical techniques such as accelerators, electron, photon, and mass spectrometers, and laser techniques, and by the many theoretical concepts and computation methods which found their way from the physical to the chemical laboratories and from quantum physics to quantum chemistry, the quantum mechanical treatments of molecular structure and reactions that revolutionized modern chemistry. Conversely, biology, but especially chemistry, aided physics to extend its domain to the more complex systems of these fields. Each branch of science is embedded in the rest especially the neighboring ones. The parts of science cross fertilize, draw from and reinforce each other; they are indebted to each other. This process is continuous and wide spread and accounts for the unceasing readjustment of the functions of science within and between its parts and the ultimate cohesiveness and advancement of science as a unified whole.

In a similar fashion, one can establish the mutual embeddedness of science and engineering. The transistor, the knowledge of the atom and its nucleus, and materials science have revolutionized modern technology, and *vice versa* engineering devices such as vacuum systems, sophisticated detectors, and computers have been the life blood of the vitality of science (Chapter 7).

The Coordinating Principle. Out of the principle of embeddedness emerges naturally the coordinating principle of science. The mutual interaction and accommodation of the parts of science are possible if an effective, continuous, and dynamic adjustment takes place in the activities of the scientist to the results obtained by his colleagues. Since such mutual adjustment depends on the independent decisions of every scientist, the operation of the coordinating principle of science requires scientific freedom.[25] The coherence of science and the cohesiveness of scientific knowledge rests on the traditional scientific values which are embodied in the dedication of the scientist to an intellectual process which ironically is virtually beyond his control. This reliance on scientific values is especially significant today when extreme specialization, restrictive training, and diverse professional goals hinder the effective coordination of the advancement of science. A little diversification, so characteristic of science's past, is thus necessary, as is a broad education of the student in science. It has been correctly said that if scientists were kept strictly without any mutual communication for a few hundred years, the total discovery achieved by them would be little more than what is normally gained by science in a few years. It is, therefore, imperative that the "coordinative functions of freedom in science"[30] are understood and respected by society and especially by the administrators of scientific institutions. Perhaps the non-scientist can appreciate this important element of the advancement of science on the grounds of social efficiency.

The Principle of Correspondence. This principle was initially applied to unify the new laws of quantum physics and relativity and the old laws of classical mechanics, the microcosmic with the macrocosmic phenomena, the events on the short-time scale with the events on the long-time scale. It requires agreement in the prediction of the laws of a more general theory (e.g., quantum physics and relativity) and the laws of the older more restricted theory (e.g., classical mechanics) in the regions the two domains overlap, and a smooth emergence from one domain to the other. The two theories must *correspond* where their realms of validity overlap. Thus relativity and classical mechanics are not contradictory, but simply the new theory (relativity) is more general, of which the old theory (classical mechanics) is a special case. The principle of correspondence requires the relativistic formulae to reduce to those of classical mechanics when objects move at relatively low speeds. The principle of correspondence would also require that quantum mechanics reduces to classical mechanics in the limit where the principal quantum number n approaches infinity.

As science progresses into new fields, it carries along the old laws and many of the old concepts. While new concepts and new quantities are introduced to describe the new phenomena, the experimental measurements and observations in the new fields are expressed -- albeit with more sophistication -- in terms of the classical concepts of energy, momentum, position, time and so on. The experiments themselves are based on the measurement of signals generated by light or charge. This is a correspondence in concept and in measurement.

The principle of correspondence provides a golden rule as to the relation of new discovery to existing knowledge in science and facilitates its effective coordination and unification. It is of great value as an instrument of guidance in discovery for it emphasizes scrutiny of the new in the light of the existing, that which is already a part of us. This referral of the new knowledge to the old knowledge is by no means hindering. On the contrary, it establishes and integrates the new and modifies and expands (or replaces) the old knowledge. It forces the scientist to seek consistency not just between his own findings, but also between the findings of everyone else current or past. The habit of this referral is essential for science. It is also essential for the rest of society, for in society too the new must be scrutinized in the light of the existing. The new must be looked upon and be assessed in relation to the endured traditions, values, and cumulative aspects of life which have been transmitted from generation to generation and have formed the standards and the norms of society.

The Principle of Complementarity. Much has been said about this principle and deservedly so. It was introduced by Niels Bohr[31] to describe the wave and particle properties of the electron (and other atomic and subatomic particles), the wave-particle duality. It recognizes that it is possible to grasp one and the same event or situation by two distinct, mutually exclusive modes of interpretation; two concepts (wave and particle in the case of the electron) both of which are necessary for a more complete description of reality, but which cannot logically coexist. The two aspects of reality -- the two conceptual ways of describing reality -- complement each other in an almost paradoxical way for neither is comprehended in, nor is it reduced to, the other.

There are many other examples of complementary situations. Let us refer to the commonly used example of the complementarity between the dynamic behavior of the motion of an individual molecule in a gas and the macroscopic behavior of the gas as a whole, between the thermodynamic and the atomistic pictures, the kinematic and the dynamic description, each of which is appropriate to a context quite different from the other. The macroscopic properties of the gas as a whole -- for instance, the gas temperature and the gas pressure -- can be understood, respectively, in terms of the average kinetic energy of its constituent molecules and in terms of the average of the forces exerted by the collision of these molecules on the surface of the gas container. This kinematic description of the macroscopic properties of the gas in terms of averages, however, says nothing about the dynamic description of the motion of each and every molecule comprising the gas. The description of the gas based on kinetic theory and the description of the gas based on the dynamics of the individual patterns of molecular behavior are complementary. We learn more about the gas and its constituent molecules by using both descriptions than by using either description alone. The dynamic aspect (microscopic) is appropriate when one is dealing with only a very few molecules and wants to understand their behavior and interactions, the thermodynamic (macroscopic) aspect is appropriate when one is dealing with a large number of particles (say, a high-pressure gas or bulk matter) and only large-scale observations about it. The former is applicable when we want to understand the interaction of one molecule with another (say, the interaction of the SO_2 molecule with the H_2O molecule in the atmosphere to form acid rain), the latter when we are interested, say, in the weather conditions that may lead to the development of storms. In the microscopic description, the dynamic properties of individual molecules are the crucial input. In the macroscopic description, it is the average properties of a large number of particles which are relevant. While the macroscopic system is in one sense made up of its microconstituents, it is, in another sense, not reducible to them.

The principle of complementarity has wider philosophical significance. There exist many aspects of, and relations in life, which are complementary -- not opposing -- and there is a need to recognize these as well as the existence of complementary approaches to many of society's problems. To quote Oppenheimer,[32] "never before have we had to understand the complementary, mutually not compatible ways of life and recognize choice between them as the only course of freedom." Thus, there exists a complementarity in the various ways to the truth, the intellectual and the material needs of man, the physical and the moral behavior of man (both been irreducible representations of his behavior), the role of science as a way to the truth and as a servant of man's material needs, the responsibility of the scientist as a researcher and as a member of society, the microscopic (individualistic) and the macroscopic (societal) view of society, the rights and the responsibilities of the individual and the society; between what we can do for and what we receive from society. There seems to be, as well, a complementarity between such concepts as security and freedom, and between the national and the international functions of individuals and governments alike. Furthermore, there is a clear complementarity between what belongs to Caesar and what belongs to God, between science and religion (Chapter 9).

In spite of the fact that often the seemingly opposing are complementary, a distinction needs to be made between the two, the complementary and the opposite. The former complement, integrate, complete the description or the essence of things or situations, the latter are contrasts. The former are "and" (particle *and* wave, scientist *and* citizen), the latter are "or" (light *or* darkness, peace *or* war, good *or* evil). In stark contrast to the complementary entities, situations or descriptions, opposites are not faces or aspects of one and the same reality although they have often been used as such. For instance, antinomic theology seeks more complete understanding by oppositions of contrary but equally true propositions.[33] We are also reminded of Plato's picture of joy and sorrow. The two, said Plato, are like two cherries hanging from the same stem. A person cannot taste one without touching the other. Perhaps the difference between complementary and opposites can best be illustrated by Niels Bohr's statement: "The opposite of a correct statement is a false statement. But the opposite of a profound truth may well be another profound truth."[34]

The Principle of Continuity. Essential as this principle is for the functioning of science, it is not an intrinsic principle of science. This is rather an operational principle referring to science planning and science policy. Scientific discovery is often liken to fire. As in the case of fire, if the scientific effort "burns well" the administrator should leave it alone; he should not intermingle with it. If the fire subsides, he should stir, but, even then only gently. The administrator's foremost duty is to secure the precious continuity in the vitality of science for this is in the best interest of both science and society. Science cannot, like a faucet, be turned on and off. Even in public works such actions are costly. The notion that whatever science is needed can be had when it is needed is as wrong as it is irresponsible. The needs of science change with time, the fertile environment in which it grows easily deteriorates, and the equipment and facilities it needs rust and quickly become obsolete.

5.3. THE SCIENTIFIC METHOD

We referred to this in chapter three. However, it is only natural that we revisit it again in view of what we have discussed in the intervening chapters, especially the present one. The persistent qualities, characteristics, and principles of science we discussed in this chapter define the way science works and advances, the so-called *scientific method.* This is an unspecified and at times an imperfect and tumultuous process which is nevertheless well characterized and dynamic. The scientific method uses the peer review system, engages the colleague as a referee and as a juror, and provides innumerable ways and forums for the free and open exposition and debate of the facts. However, not all of science is done this way. Indeed, many have argued that there is no such thing as "the scientific method." Nonetheless, the method of science *is* the ways the scientists act, and the progress of science is much more than their individual accomplishments. At any given time, a point of view prevails in science from which derives the elementary creative act of scientific work, namely, the elemental choice between alternatives and perceptive planning. It is within this predominantly accepted scientific view that the scientist and the science worker conduct their vocation. Facts or theories contradicting this accepted view are treated with caution and are often met with resistance.

A most distinct characteristic component of the scientific method is the tacit involvement of the scientist and the science worker in the acquisition and understanding of new knowledge. Their intuitive judgements and subjective feelings of "reasonableness" cannot be separated from findings eventually published by them as has been correctly pointed out by Polanyi and others. Nor can it be denied that the scientist accepts concepts and ideas even when they are contradicted by the available data at the time. The fundamental element about the way science gets at the truth is not the correctness with which the scientist comes to conclusions, but that since he makes mistakes publicly they are correctable.

Science's discoveries, devices, principles and free ways to the "truth" instill in its practitioners a way and an approach to the secrets of nature which is characterized by the values, the principles, and the persistent qualities of science that constitute the roots of the scientific tradition. In the last analysis scientific discovery is an art for which it is impossible to lay down the rules. To quote A. D. Ritchie,[35] "It is quite impossible to lay down rules knowing which anybody can write poems like Shelley or make statuses like Praxiteles. So also is it impossible to lay down rules which will enable anybody to make discoveries like Faraday or Pasteur."

5.4. REFERENCES AND NOTES

1. A. N. Whitehead, *Science and the Modern World*, The MacMillan Company, New York, 1926, p. 298.
2. W. Heisenberg, Daedalus LXXXVII, Summer 1958, p. 95.
3. E. Segré, *From Falling Bodies to Radiowaves*, W. H. Freeman and Company, New York, 1984, p. 164.
4. This, in spite of the fact that physical laws are limited human constructs (see Section 4.4 and later in this Chapter).
5. M. Polanyi, *The Study of Man*, The University of Chicago Press, 1959, p. 12.
6. J. R. Oppenheimer (Science 111, 14 April 1950, p. 373) cited a letter by Thomas Jefferson in which he wrote: "Science can never be retrograde; what is once acquired of real knowledge can never be lost."
7. G. Holton, in *Science, Technology, and Society*, R. Chalk (Ed.), American Association for the Advancement of Science, Washington, D. C., 1988, p. 48.
8. W. Heisenberg, *Physics and Beyond*, Harber & Row, Publishers, New York, 1971, p. 98.
9. Ref. 8, p. 195.
10. J. Bronowski, *Science and Human Values*, Harper & Row, Publishers, New York, 1965. p.14.
11. A. D. Ritchie, *The Scientific Method*, Harcourt, Brace & Company, Inc., New York, 1923, pp. 42-43.
12. I. I. Rabi, *Science: The Center of Culture*, The World Publishing Co., New York, 1970, p. 41.
13. H. Mohr, Angew. Chem. Int. Ed. Engl. **17**, 670 (1978).
14. B. R. Russell, *Religion and Science*, Oxford University Press, London, 1960, p. 169.
15. V. F. Weisskopf, Science **176**, 14 April 1972, p. 138.
16. A. L. Schawlow, quoted in D. C. O'shea, W. R. Callen, and W. T. Rhodes, *Introduction to Lasers and their Applications*, Addison-Wesley Publishing Co., Reading, MA, 1978, p.1.

17. Th. S. Kuhn, *The Structure of Scientific Revolutions*, The University of Chicago Press, Chicago IL, 1970.
18. J. R. Oppenheimer, *Science and the Common Understanding*, Simon and Schuster, New York, 1954, p. 24.
19. J. R. Oppenheimer, *The Open Mind*, Simon and Schuster, New York, 1955, p. 114.
20. H. D. Smyth, American Scientist XXXVIII, July 1950, p. 426.
21. One need be reminded of the fate of genetics during the Lysenko years in Stalinist Russia in modern times.
22. G. Gorelik, Physics Today, January 1996, p. 61.
23. American Scientist, *The Integrity of Science; A Report by the AAAS on Science in the Promotion of Human Welfare*, **53**, 174 (1965).
24. H. T. Davis, *Philosophy and Modern Science*, The Principia Press, Bloomington, Indiana, 1931, p. 3.
25. M. Polanyi, Minerva V, Summer 1967, p. 533.
26. R. Seindenberg, *Anatomy of the Future*, The University of North Carolina Press, Chapel Hill, 1961, p. 8.
27. R. Feynman, *The Character of Physical Law*, The M.I.T. Press, Cambridge, MA, 1967, p. 173.
28. Italics are the author's, not Feynman's.
29. A. M. Weinberg, Bulletin of the Atomic Scientists XXII (No. 4), April 1966, p. 8.
30. M. Polanyi, in *Physical Science and Human Values*, Princeton University Press, New Jersey, 1947, p. 124.
31. See for instance, H. J. Folse, *The Philosophy of Niels Bohr*, North-Holland, Amsterdam, 1985 and A. Pais, *Niels Bohr's Times*, Clarendon Press, Oxford, 1991.
32. J. R. Oppenheimer, *The Open Mind*, Simon and Schuster, New York, 1955, p. 144.
33. V. Lossky, *In the Image and Likeness of God*, St. Vladimir's Seminary Press, Crestwood, New York, 1985, pp. 51-53.
34. W. Heisenberg, *Physics and Beyond*, Harber & Row, Publishers, New York, 1971, p. 102
35. A. D. Ritchie, *The Scientific Method*, Harcourt, Brace & Company, Inc., New York, 1923, p. 53.

CHAPTER 6

The Scientist
and
the Science Worker

Who are the people who do and work in science? If *science* is no longer as distinct a term as it once was, is there still distinctiveness in the term *scientist*? And if, as Boris Pasternak says in Dr. Zhivago, gregariousness and mass mentality are the refuge of mediocrity what can be said of the huge numbers of those working in science today? I shall attempt to answer these questions not as a historian of science, but as a working scientist.

There are many ways to view and to work in science and many functions to be performed. There are those who view science widely as an airplane passenger views the land beneath the plane, and those who view science closely and warmly as the citizens of an old city who constantly walk its narrow streets. There are those who work in the busy, quiet, and often forgotten places called laboratories, those who barely bother to know what goes on in such places, and those still who transform science's laboratories into offices for paper shufflers and regulators. Those who dig deep for fresh water, and those who build dams and channel the waters which flow on the surface. Those who influence the crops of knowledge to be cultivated, and those who select the products which are transmitted to the market (to be published). There are still those who make discernible paths through the jungle of the unknown, those who come after them to build highways on their paths, and even those who drive along the highway enjoying the smooth ride often disrespectful of the efforts that have led to its construction. I speak of the *pioneers,* of the *settlers,* and of the *actors.* Those who do science, those who use science, and those who speak about and manipulate science. I shall, however, restrict myself to the scientist and the science worker. Both are indispensable elements of science, but there is a need for a distinction. *Not everyone who works in science is a scientist.* There are in science today many science workers and relatively fewer scientists.

6.1. THE SCIENTIST

Who is he? He is the *heart* of science because science has been and still largely remains a particularly individualistic profession. He is privileged to the secrets and the beauty of nature and a witness of the endless waves of scientific thought and discovery. He does science because he simply assumes that there must be rational answers to his questions. His work is thus underlined by a simple article of faith: *nature is rational.* For how can he understand rationally the irrational? "Over the entrance to the gates of the temple of science are written the words: *Ye must have faith,*" proclaimed Max Planck.[1] The scientist has faith in rationality, but he has no monopoly on rational thought.

A scientist adheres to the scientific ethic and to the norms of science. He is dedicated to the understanding of the universe through systematic inquiry and is responsible for the integrity, coherence, vigor, and freedom of and in science. He is the defender of the facts and the truth as they are revealed by the method of science. This dedication and this commitment the scientist learns by proper education and by proper example in the execution of basic research. It seems safe to assume that there are as many ways to do science as there are scientists, and yet the ways the scientists execute their vocation constitute the spring of tradition in science and is an indispensable element of the soundness of scientific knowledge. They provide for the scientist patterns of thought and examples of practice of his vocation and a style of life full of excitement and worthiness. The qualities of a scientist are rare and difficult to achieve and maintain, and they are never all concentrated on a single person. Neither these qualities, nor the technical knowledge itself make "wise men from foolish or good from wicked."[2] Nonetheless, it is through them that the scientist pulls science and society in new directions in which man and society eventually equilibrate.

As People They Are as Different as People Can Be. Scientists vary in appearance and speech, education and culture, manners and civility, political and social views, philosophy and beliefs. Some tend to be outspoken and straightforward in their views albeit often introverted, shy, idealistic, and inarticulate. They are generally open-minded, liberal in their attitudes and independent, and yet they are analytical, convinced of and persistent in their positions and often submit with relative ease to national and institutional authority. They are aggressively ambitious[3] and this at times makes them become prisoners of power and money for they need support to do what they passionately love to do: science. They are tolerant of ambiguity while they seek certainty. They believe, as James Gleick wrote of Richard Feynman, "in the primacy of doubt not as a blemish upon our ability to know but as the essence of knowing."[4]

They are intellectually rich and intellectually narrow. In the opinion of René Dubois, "the intellectual narrowness of many specialists comes from the widespread assumption that the discovery of new facts is the most important aspect of knowledge."[5] Some are interested in art, music, literature, philosophy, and history and yet others never developed such interests. For instance, according to Hovis and Kragh,[6] Paul Dirac never developed an interest in art, music, or literature, and he seldom went to the theater. Among them are philosophers, idealists, stoics, peripatetics, social advocates,[7] along with down-to-earth pragmatists, and egomaniacs. One finds among them persons of modesty,[8] grace and generosity who inspire respect and emulate imitation, and persons uninteresting, mean spirited, tart and feared critics. One finds among them also persons of sacrifice and persons of self interest, persons who help others and persons who are jealous of colleagues and protegees.[9] Extremely rarely one encounters persons who take advantage of those they had helped.[10] With rare exceptions, they like honors and fancy titles and are flattered by them. They know to rub shoulders, promote and be promoted, nominate and be nominated for awards, positions, committees, councils, academies and the like.[11] Still others enjoy their role as celebrity and in this role some had immunity from criticism.[12] In their ranks are persons of extreme virtue and rarely persons of extreme vice.

They come from all kinds of family and national backgrounds. Some are nurtured in the tenderness of loving homes, others bare the scars of unhappy childhoods. Most have happy and beautiful families themselves, others fail, remain aloof or are detached of family. Some blossom early and others blossom late giving little early indication of their talents and mental powers. Distinct examples of this late blossoming are Newton, Darwin, and Einstein.

Some are deeply religious (e.g., G. Galileo, I. Newton, M. Faraday, A. -M. Ampère, A. J. Fresnel, E. Rutherford, see also Chapter 9), others are agnostic and atheists. Still others (e.g., M. Planck, A. Einstein) look at God through the beauty of the laws of nature, are preachers (e. g., M. Faraday[13] and C. A. Coulson), Sunday School teachers (e.g., H. Eyring),[14] search for meaning beyond the confines of science into mysticism (e.g., E. Wigner, W. Pauli, R. Oppenheimer), or exceed the bounds of science and make religious-like antireligious proclamations.[15]

Although the spirit of science is antipodal to the spirit of war, there has been an intimate connection between science and war going back to Archimedes, Leonardo da Vinci, Galileo, and the French revolution.[16] Time after time scientists served the military needs of their warring nations. In WWI Marie Curie worked at military field hospitals and brought to the French WWI wounded the diagnostic help of X-rays.[17] H. W. Geiger and E. Marsden were at the Western front in opposing camps; H. G. J. Moseley was killed in the battlefield, E. Schroedinger served as artillery officer; J. J. Thomson, W. Bragg, W. Crookes, O. J. Lodge, and E. Rutherford participated in the British war effort; and over 3000 German professors (including E. Fischer, P. Lenard, F. W. Ostwald, M. Planck, W. C. Roentgen, W. Wien, H. W. Nerst, F. Haber, and F. Klein) signed the German manifesto, published in October 1914, which defended certain acts of that war.[18] Similarly in WWII, the scientists of warring nations, overtaken by patriotic justification for their actions and attracted by the excitement and the "absorbing" scientific challenges of the new armaments,[19,20] immersed themselves willingly and passionately in weaponry and developed (in both the West and the East), the atomic bomb and later the hydrogen and other types of bombs. "In the spring of 1942," recounts[19] E. Teller,"both Fermi and I were at Columbia University. Physics had moved closer to the grim realities of war. Many of us had started to work on the fission bombs. .... It is hard to describe the intensity and the fascination of the discussion that followed. ..... The spirit of spontaneity, adventure, and surprise of those weeks in Berkeley was never recaptured for me in the many years of hard work in which atomic bombs were developed. "The same excitement was shared by the counterparts in the former Soviet Union. Igor Tamm, Andrei Sakharov and their colleagues were "working on the (atomic bomb) project with tunnel vision, fully absorbed in the atomic problem."[21]

There were scientists who urged the politicians to build the atomic bomb (notably A. Einstein and E. Fermi in the United States),[22] scientists who built the atomic bombs, and scientists who refused to do so (such as Wolfgang Pauli[23] and Lise Meitner[24] in the West and Peter Kapitsa in the East[25]); scientists who signed petitions during WWII against introduction of the bomb in the war against Japan,[23,26] scientists (e.g., John von Neumann[27]) who after WWII advocated that the bomb should be used to wipe out the Soviet Union before it developed its own atomic bomb, and scientists who helped build the atomic bomb and were later remorseful and regretful; their anguish had been encapsulated in the infamous statement by Robert Oppenheimer after Hiroshima and Nagasaki: "the physicists have known sin."[28] And there are still scientists who stopped working on further development of atomic weapons after they had helped built the first atomic bombs (e.g., Victor Weisskopf),[29] scientists who never stopped improving the means of mass destruction (such as the archetypical scientist-warrior Edward Teller), scientists who are involved in armament build up and in armament limitation or disarmament, scientists who served their country, and scientists who spied against their country (e.g., K. Fuchs secretly gave Russia the scientific results of his American colleagues to develop the atomic bomb). In his book *The Joy of Insight*,[23] Weisskopf candidly expressed his thoughts regarding his decision to join the Manhattan project in 1943:

> "Today (1991) I am not quite sure whether my decision to participate in this awesome -- and awful -- enterprise was solely based on the fear of the Nazis beating us to it (the atomic bomb). It may have been more simply an urge to participate in the important work my friends and colleagues were doing. There was certainly a feeling of pride in being part of a unique and sensational enterprise. Also, this was a chance to show the world how powerful, important, and pragmatic the esoteric science of nuclear physics could be."

Scientists like and dislike their colleagues. They are charitable to their competitors and mean spirited toward their competitors.[30] They rejoice in their success and at times at their failure such as when James Watson and Francis Crick learned that their rival colleague Linus Pauling guessed the wrong structure of DNA.[31] They fight over priority and credit and become deeply unhappy in the process just as did Philip Lenard in connection with the discovery of X-rays,[32] and the discoverers of penicillin.[33] They often resist unorthodox ideas and profound new discoveries,[34] and even block colleagues from exposing their views when they are in conflict with their own.[35] They wound, they disappoint, and they embitter.[36] And in this, too, they are no different from persons in other walks of life, except in one singularly significant way: the scientist can hurt professionally his colleague far more easily and irrevocably than persons of other vocations because the scientist relies heavily on his colleagues not only for their contributions to science and for information, but also for credit and fair criticism. A scientist's work is confronted by and is integrated with that of his colleagues and a scientist's colleagues are his collaborators, his competitors, his peers, and the judges of his work. *They* determine the value of his contributions to science and they are entrusted as the competent group to judge his work. A scientist can thus be immeasurably hurt by the malicious or the careless colleague. It appears that this intimacy and forced closeness between the scientists is both a cherished and a beautiful experience, but at the same time it is a source of strain between the scientist and his colleagues. Criticism is a fundamental element of the business of doing science and, with rare exceptions, scientists welcome constructive criticism of their work, but they react adversely to the malicious criticism. Some of the great scientists (e.g., Planck) responded to criticism positively and improved their work. Others (e.g., Kepler, Newton) did not easily accept criticism, and still others got depressed because of it (e.g., Boltzmann).

No person is without sin, and the scientist -- as scientist -- is no exception. Under the stress of enormous pressures from within science and from outside science, the scientist often conducted himself in regrettable ways (see Section 6.4) that diminished his stature. The scientist's new roles, implications of his work, and involvement with politics and money affected his spirit, influenced his attitude, turned black and white areas into gray, and made him skillful in knowing -- as the mysterious priestess Pythia of Delphi -- how to avoid telling the truth without saying a lie. Nonetheless, as Snow[37] noted, "The remarkable thing is not the handful of scientists who deviate from the search for truth (and the norms of science) but the overwhelming numbers who keep it." This, as he said, is a demonstration of moral behavior on a very large scale.

As Researchers. To be a scientist is to do research and to do research means to publish. It is the business of the researcher to obtain new information, develop new understanding, discover a new "law," and do something for the first time in the history of humanity. In this the scientists have their research styles and approaches and they are perpetually moving hunters. They are guided as much by intuition[38] as by vigorous logic, pursue relentlessly anomalies,[39] skillfully device traps for nature to reveal itself, and possess a unique ability to improvise. They practice science as individuals and intervene with their experimental method into nature through a myriad of disciplined ways aiming at measurable and interpretable results. Some believe in the supremacy of mathematics (e.g., Poincaré, Dirac, Pauli, Hawking, some modern theorists), and others triumphed using *no* mathematics at all (e.g., Faraday). Some are drifters, jumping from one fashionable field of science to another, others are skillful myopic mine diggers in narrow scientific fields, and yet others are superficially visionary attempting to make theories of everything. Some are thorough, comprehensive, perfectionists who repeat the same measurement over and over again, others are content with the first-order (back of the envelop) approach, are tolerant of greater degree of uncertainty and are satisfied with fewer measurements. They all speak in a coded language -- alien to most people -- and they all make mistakes.

They hold sacred their responsibility to truthfulness in their factual information for they know that the objectivity of science depends on it. They know that the advancement of science through scientific research is first and foremost based on trust and that fraud in science is not just wasteful, but also corrosive. Nevertheless, there have been isolated cases of data suppression. A recent case that can be cited involves the carcinogenicity of vinyl chloride used in the plastics industry where there is evidence[40] that "a considerable number of scientists were aware of hazards of vinyl chloride long before the facts were made available to the NIOSH (US National Institute of Occupational Safety and Health) or the public; yet they kept quiet and gave no warning." A second similar case has been unfolding for some time involving scientists in the tobacco industry and their role in suppressing the facts as to the connection between cigarette smoking and cancer. In isolated instances practitioners of science were discovered publishing results that were fictitious. In 1974, the magazine "Science" announced[41] "A young investigator at the Sloan-Kettering Institute for Cancer Research in New York, is alleged to have falsified the results of an experiment intended to prove that skin, when grown in tissue culture, loses its ability to provoke an immune response." The same magazine in 1981 reported[42] that in 1980 alone "Four major cases of cheating in biomedical research have surfaced." Others (e.g., Broad and Wade[43] and Brush[44]) used these incidents to attack science and scientists in earnest.

Scientists recognize the importance of publication and they have a "habit of covering their tracks" when they publish the findings of their research. Only the successes are normally reported by them and only the successes are normally remembered. How many geometrical shapes for the planetary orbits Kepler employed before he came to the ellipses? How many paths Planck followed in his struggle to comprehend the nature of blackbody radiation before he assumed quantization? We will never know. In the words of Faraday "The world little knows of how many of the thoughts and theories which passed through the mind of the scientific investigator have been crushed in silence and secrecy by his own severe criticism and adverse examination. How many of his preliminary ideas did not make it."[45]

The active scientist is deeply interested in his research. He is anxious to solve his specific problem, and he is fascinated when he finds a solution to it. In search for soundness of his findings he ruthlessly questions himself and without mercy he tries every conceivable way to prove to himself that he is not wrong, constantly thinking of alternatives and possible causes of error. The creative scientist is an imaginative, original, and self-disciplined thinker. His mind is alert, his art is precise, and his approach is pragmatic. With enthusiasm he immerses himself deeper and deeper into his research, learning as he works, reflecting and readjusting as he discovers, becoming inseparable from his work, constantly challenging himself from one level of discovery to the next. His work becomes an extension of himself just like music is an extension of the singer's feelings and the violin an extension to the gypsy's emotions. He usually has a large capacity for work and struggles to keep abreast of the larger pattern of things, heading for the frontier borders of his field of expertise and the hybrid disciplines. He is caught up with his work often beyond his strength. He believes in what he is doing and courageously defends it as "unique," "urgent," "important," "promising." Yet, he has every reason to be humble: he knows that his discoveries can be accepted only if they can be verified by others and that what he discovered could and would have been discovered by someone else. He is cognizant of the help he had received along the way from colleagues and collaborators. Nonetheless, a sense of due satisfaction penetrates his bones because each contribution -- "his contribution" -- in science is unique. This satisfaction becomes especially complete when his work is cited for its value by his colleagues. This holds true for both the pioneers and the settlers and even the repairers who fill the holes on the highways, beautify their sides, strengthen and smooth their rough edges. Collectively they build and maintain open for everyone the transitory avenues to and through the unknown, fully aware that these avenues will in time be improved, broadened, changed in direction, and be superseded.

While great scientists often work alone, the scientist's style is, especially recently, collaborative. They form their "research families," groups of colleagues and graduate students held together by a cohesion of mutual interest, common goals, respect, sacrifice, and sheer excitement, especially when such groups are in the process of major discoveries or radical conceptualization of a particular field. The traditional coherence of such groups centers around a leader who serves as a scientific inspirational model and who is actively emulated, and requires sustainable resources and institutional commitments. However successful such units are, they are transient. Their lifetimes are typically 10 to 20 years. Often they grow big, become diluted and fission. Sometimes these wonderful intellectual and social units die because of their successes. Regrettably, but understandably, many research families sicken by the virus of individual interest and ambition and become infected by jealousy and conflict which inflict their eventual dissolution and the accompanied unhappiness and remorse so reminiscent of the breaking up of a person's family. Though transitory, the bonds, the excitement, and the memories of the research family never perish. The research families are especially beneficial for the graduate students who are the "children" of the research family, so necessary for the fulfillment and continuation of scientific life. More often than not a scientist's research family embraces the whole world, transcends national bonds and boundaries, is cosmopolitan.

As Teachers, Communicators, and Mentors. The scientist's urge to discover is often, but not always, matched by his eagerness to tell through teaching about the new knowledge and truths he found and the tortuous ways which led him to them, their potential value and limitation. Many feel it as part of their responsibility to educate and to motivate the young. However, not all good scientists are good teachers or good lecturers. But when they are, it is a real joy to listen to them, to absorb their words as their thoughts unfold and their arguments develop. Richard Feynman is considered one of the recent fine lecturers in physics, but the classic example here is that of Faraday. Many scientists have difficulty expressing themselves or their thoughts.[46]

It is also the responsibility of the scientist to work with students. To bring them to the furthermost frontiers of science and to let them take the next step, and, in practicing science together, to teach them the norms of science tacitly. Unfortunately, the joy and excitement of working with students is not universally shared among scientists and many scientists, including some of the greatest (for instance, Newton, Faraday, Maxwell, Einstein, Dirac) had no students. And yet some were themselves fortunate to have had good mentors (e.g., Newton had Isaac Barrow and Faraday had Humphry Davy) through whom they got support, encouragement, direction, early exposure, and contacts. On the other hand, others (such as Rutherford and Bohr) were successful mentors and produced large numbers of outstanding new generations of scientists. Haskins[47] stressed the importance of such scientific enclaves -- centers of excellence -- and gave a beautiful example of their value: Hans A. Krebs (1953 Nobel in Medicine and Physiology) worked with Otto H. Warburg (1931 Nobel in Medicine and Physiology), who was associated with Emil Fischer (1902 Nobel in Chemistry), who worked with Adolf von Baeyer (1905 Nobel in Chemistry). The genealogy extends beyond the era of the Nobels: von Baeyer was a student of August Kekulé (carbon compounds, aromatics), who studied with Liebig (foundations of modern organic chemistry), who worked with Gay-Lussac (fundamental laws of the behavior of gases), who was a student of Berthollet, who in turn was a student of Lavoisier.

Their Common Characteristics. Scientists are persons with
- great mental power, sustained thought, imagination, curiosity, and creativity;
- sound judgement, critical spirit, intuition, and resourcefulness;
- initiative, risk taking, courage, and perseverance;
- passion for and emotional involvement with their work;
- deep knowledge of the foundations and frontiers of their field;
- a shared value system based on freedom of expression and respect for truth.

These characteristics are difficult to achieve and maintain. They are neither all concentrated on a single person, nor are they all concentrated on every scientist to the same degree. These characteristics along with the principles which underpin the ways of action of the scientist (see Chapter 5) constitute the spring of knowledge and tradition in science and are an indispensable element for the soundness of scientific knowledge. They provide guidance for the scientist's work and worthiness for his life and account for his dedication to his vocation. Of the inner beauty of this rather loosely defined vocation the non-scientist knows nothing and the science worker only very little. It is nonetheless qualities such as these that restrict the scientist's numbers, separate him from the science worker, and require society to respect the ways he conducts his work.

6.2. THE SCIENCE WORKER

It is often pointed out that the majority of all who ever worked in science are working in science today. The proliferation in the numbers of those working in science, the expansion of the scientific endeavor, the vast uses and applications of scientific knowledge, and the cost of modern science, have resulted in more than one scientific identity. Not just more, but very different people are attracted to science today.

Today, the term scientist is used loosely embracing a group of people with a broad and heterogeneous spectrum of expertise, interests, and attitudes. The term used in this loose manner tells us virtually nothing about their values and motivations. As we discussed in the preceding section, although many of the qualities of the scientist are widely spread, they are normally concentrated on the very few, the scientists. The rest of those working in science we shall refer to as the *science workers*. A science worker is an important, but is not the crucial element of the advancement of science.

The science worker is normally indifferent in matters of principle. His interests are more subtle, often in support of the scientist. As opposed to the scientist, the science worker may have problems in temporary self limitation and he may not be "driven" to get to the heart of things.

The spectrum of science workers is wide. It includes many former scientists who slid into other roles, many skillful experts, many with enormous craftsmanship or encyclopedic knowledge or excellent entrepreneurial abilities, many articulate spokesmen of science with interests in politics and management, many brilliant administrators who make serious decisions affecting the functioning and the future of science, and even many directors of large research and funding institutions and presidents of academic and industrial organizations. Among them are also many inventors, but few discoverers.

6.3. Their New Environs and Changing Functions

"How does one become accepted as a professional scientist?" asked Thomas Kuhn.[48] His answer was: by professing in a paradigm, that is by accepting a way of looking at a set of facts which the paradigm itself selects as significant. There is more to becoming a scientist than that. Indeed, Polanyi[49,50] goes deeper. First, he says, one has to have a love of science and a faith in its great significance which precedes any real understanding of it. Second, he says, he has to surrender to the intellectual authority of science. Third, he says, as he becomes a professional scientist, he must listen to his scientific conscience. To love and to have faith in science one must first have some understanding of science. How then can one love and have faith in science before he develops an understanding of it? And how can one surrender to the intellectual authority of science (profess in a paradigm) and listen to his scientific and human conscience? The answer must be: through proper education and through proper training.

Their Education and Training. We shall discuss education in Chapter 8 and for this reason we shall be brief here. The education and training of a scientist today is a long and expensive process, much more than in years past. For this reason many argue that it should be reserved for the capable and the motivated. In years past, the student in science was educated more and was trained less. The reverse is true today. The many specialized tasks and the new environs of and mechanisms for scientific life seriously influence the temperament and motivation of the student. Most students today demand, and most universities today provide, training not education. Even though the average student today has a great deal more specialized knowledge than in the past this is too little in relative terms. A narrow and directed training inevitably limits and leads to eventual incompetence and obsolescence. The increased demand for training and the squeezing out of education endangers the future of both science and society and handicaps the student. Specialization must be superposed on education, not take the place of education.

Thus the teaching of science must emphasize the unity and the universality of science and it must become broader than the mere attempt to produce experts in specialized fields. The student must learn by personal involvement in research the norms of science and the functions of the "scientific method" and with the tender care of the individual professor he must make the transition from the classroom to the world of living science. He should be exposed to literature, history, broader cultural values and human activities, and he should be encouraged to develop a philosophical and social perspective. Education would help him recognize that beyond discovery lies the search for discovery, above the brilliant thought lies the thinking of the thinker and the thinker himself, that man and society transcend the scientist and science and that the latter must serve the former.

And, yes, the history of science is necessary for the scientist's education. Early in his career the student in science needs to develop a deep sense of the role of history in discovery and feel the urge to connect to the intellectual tradition of science especially of his own discipline. The student in science needs to connect to his scientific heritage, reflect over the virtues and deficiencies of its heroes, and learn the way of the masters by reading some of their papers. Knowing his scientific ancestry will enable him to better understand his present scientific family and establish stronger bonds with it. I am certain that he will better appreciate the statement I made earlier, namely, that science is first and foremost a human activity. He will find out for himself that in science too we stand on the shoulders of giants who have been wrong (when, for example, they regarded the world flat or when they assumed that the sun revolves about the earth, or when they held that the electron and the proton were the only elementary particles, or when they were convinced that life was generated spontaneously from inanimate matter). He will realize that judgement goes with instinct and knowledge with full submission to the facts. He will discern that important discoveries often come through signals barely above the noise and that the fine whispering of nature -- signaling new discoveries -- is often missed by rough calculations or crude measurements. In this way, then, he should find out for himself whether he aspires to become a scientist or a science worker and that in either case he has to assume responsibility for his actions.

Their Colleagues and Their Scientific Community. The scientist and the science worker have close and distant associates, colleagues from different lands whom they probably first met in the annals of scientific literature and at scientific conferences and then in science's laboratories and in the scientists' homes. They share information and understanding and they benefit from the work of each other. They even tune remarkably quickly and universally to the trends and the fashions in science.

The science community (comprising all the scientists and science workers in the world) is one of uncertain boundaries but with internal structure. It is, as Polanyi put it, an unofficial establishment based on fiercely competitive scientific excellence held together by stable, direct, and interpersonal relationships. Its five-century long tradition of communication and cooperation lies at the very foundation of science. Through it scientists advance science and understand each other. The scientist is brought up to believe that he has deep responsibility to the scientific community.

The social organization of science has its own hierarchy and is designed to foster the production of an elite in which prestige comes on the merits of work and position in the scientific community. It is an inhomogeneous organization within which often develop separate homogeneous patterns of social interactions and shared-idea systems. Such patterns restrict the scientist's objectivity and can lead to chauvinistic-kind of scientific cliques patronizing their own at the exclusion of others. Prevailing patterns of specialization, methodical conceptions connected with disciplinary homogeneity, and the absence of one of the scientist's strongest norms -- open-mindedness -- can limit a scientist's judgement.

The aims of the institutions of science and scientific societies are presumed to be the maintenance of freedom and advancement of science, the securing of support for science, and the facilitation of the means for science to best serve society. As is normally the case with other human institutions, the institutions of science tend to serve interests of their own, yield to special interests, and bow to political pressures. However, it appears that compared to other human institutions, the institutions of science are still less powerful and more decentralized. The scientific community as a whole is still largely preserving a style of behavior distinct unto itself with a great deal of the traditional norms despite of continuing concentration of power in certain "decision making" groups and favored research and academic centers.

Their Working Places. A significant condition for the advancement of science is the existence of a scientific *loci,* an intellectual local environment which is conducive to creativity. A scientific environment committed to freedom and intellectual achievement, and institutions which recognize and respect the values of the discipline-oriented professional and sooth the inescapable tension between these values and the more utilitarian and opportunistic goals of the institution. Regrettably, institutions often fail to recognize the limits of competitiveness and the perils of pressured-environments. In their efforts to cultivate group spirit and effect group activity, they diminish the value of the "principal investigator," tilt toward the science worker, and become egalitarian and thus mediocre. Regrettably also institutions partly because of the pressure of managements and their "please-the-customer" mentality shift their focus from emphasizing respect, trust, and nurturing of excellence to the meaningless gimmicks called "measures of performance" and "management by objective" that alienate their scientists and irritate their most creative staff. They surrender a great deal of institutional autonomy, twist institutional goals, and not infrequently end up cold places whose most distinct characteristic is the *absence of trust of the individual,* including the scientist himself. They follow blindly and unscrutinizingly regulations that strangle their operation and they rely on the public relations experts for their image. Consequently, scientific institutions find it difficult to make long-range commitments to worthy scientists and to important problems. Worse still, they lose the loyalty of their scientists. They are not meaningfully engaged in "participatory leadership." Their dual structures [scientific (research) ladder and administrative (managerial) ladder] have created two distinct monocultures, -- that of the researcher and that of the manager -- separated, polarized, insulated, unequal. And yet excellence and accomplishment require strong and continuous coupling between those in the two ladders at all levels.

Many of today's scientific institutions are incapable of fulfilling their obligations to their scientists and are thus responsible for wasting talent and resources. To these sad conditions might have contributed the fact that the laboratories and the institutions of science have been progressively headed by settlers and non-scientists rather than by pioneers. And yet, it is the scientists who do science, not the administrators; the latter must not plan the *doing* of science.

Their Nations. Science is universal. Its growth is rooted in the discoveries of all nations. Every generation of all mankind builds on all prior accomplishments. However, most developing and all underdeveloped countries of the world are invariably countries with little or no science.

Virtually all of science is done in the technically advanced countries. Actually, about 95 % of the world's R&D is conducted by about 20 % of the technically advanced peoples.[51] Just six countries are responsible for about 90 % of the world R&D and perhaps a larger percentage of the best science. Tragically, most of the meager 5% of the world R&D controlled by the developing countries is spent on armament research. Regrettably then, 75 % of the world's people live in a desert as far as scientific research is concerned. Rich countries become richer and poor countries become poorer in science, as they do in technology. One of the consequences of this situation is the fact that of the 90 % of the world's scientists and science workers are in the rich industrialized nations. A large number of them left their scientifically impoverished countries and found refuge in the scientifically-rich nations of the world. They change, adapt, speak with varied and distinct accents, and many retain the bonds with and memories of the cultures of their upbringing. Characteristically, Eugene Wigner was quoted[52] as having said "after 60 years in the United States, I am still more Hungarian than American. I speak English with a thick accent. Much of the (U.S.A.) culture escapes me." Nostalgically, they look back to the country they left behind. But without proper local conditions for them to be useful "at home," they stay away from their motherlands, permanent immigrants, shining gypsies and entertaining fiddlers from every land singing and dancing at the international fares of science hosted by the rich. The art of their music and the skill of their dancing make them and their motherlands proud and simultaneously sad for they both suffer by the separation.

Thus, the poor countries export not only their valuable raw materials to the scientifically and technologically rich countries, but also their priceless human resource. While the immigrant scientist and science worker is perhaps the poor nation's most significant contribution to world's science and technology, it deprives the poor country of the able and the creative, of its scientists, science workers, and engineers. The permanent migration of scientists from the scientifically poor to the scientifically rich countries can be replaced by a two-way migration. This is a difficult change to instigate because it requires indigenous scientific and technological programs and facilities which are not easy to create. But if science is to become truly the responsibility of every nation, every nation must plan and must be helped to sustain a healthy indigenous research in science, especially in the physical, chemical, and life sciences, focusing on "Little Science" and leaving "Big Science" to the rich nations. Such activity is needed to educate and train the scientist, the science worker, the engineer, and the technician *in loco.* Training abroad is expensive and risky. Most of the trainees go for the attractive and the fashionable areas of science, cut themselves off and stay away from the homeland. Many pointed out correctly that bringing bright students to the universities of the scientifically advanced countries and training them in esoteric subjects is not the solution even if a good portion of the trainees return home. For science to be wanted by a small nation, it must contribute to its economic and social development, and thus science must accept the challenge of development and be pragmatic in its aims and promises whether they refer to goods and services, improved living conditions, uplifting of national "prestige," or elevating the local overall intellectual and civil conditions.

Indigenous science and technological development are also necessary to avoid economic and cultural neocolonialism. The dependence of the small country on the science and the technology of the rich on the one hand, and the struggle and fierce competition of the rich countries for the markets and the raw materials of the poor countries on the other hand, rekindle the dreadful fears of colonialism and its brutal exploitations. Science and technology (alas!) should have no part in any such thing. And yet aren't coal, iron, oil, and other raw materials the basis of not only the rich nation's wealth but also the rich nation's hold on power and control? Aren't the acquisition, use, and exploitation of all these resources in the hands of multinational corporations all of which employ and deploy the latest in scientific and technological advancements to enhance their grab of larger portions of the world and its resources? Science intensifies and accelerates these competitive factors because the nations with scientific knowledge know where to find resources, how to extract resources, how to best use resources, and how to control information in this long process. Unavoidably, this aggravates and inflames ideological conflicts and breeds the threat of war. Clearly, then, *it is not the luxury of science in small countries that man cannot afford, but rather the consequences of the scientific and technological isolation of most of humanity.*

Indigenous science is necessary for the efficient use of a nation's resources and by extension of the resources of the earth. A nation which borrows its basic knowledge is handicapped in its efforts to innovate and to develop, unable to profit from imported technologies. Internal scientific growth and maturity are the key to development, for even in science and technology *to be able to receive a nation has to be in position to give.*

For science, then, to be the responsibility of every nation, every nation's science should be the responsibility of science. To date the greatest part of the international dimension of science is provided by individual scientists through contacts, scientific exchanges and common projects, and by the institutions of science, scientific societies, international research centers, organizations, and foundations. But much more is needed and the burden is heavy. It lies on the shoulders of every scientist, every institution of science, every organization, and every government.

Regrettable Actions. The modern environs influenced negatively the conduct of the scientist in many and distinct ways. We illustrate these developments by a few generic examples.

- *Scientists abdicated their responsibility to the norms of science.* Whether in a free or in a totalitarian state, the scientist's role has been heavily dictated by the conception of the national interest. A fraction of the scientists and a larger fraction of the science workers had been and are being asked to virtually become "soldiers not in a uniform."[37] In this capacity they are called upon to out think adversaries, develop defenses against all possible enemies, and serve as an intellectual resource for every eventuality. In this process they often abdicate their scientific norms and comply with the demands of national security and secrecy. Most of this began with WWII. The scientists, many believe, erred when they encouraged their governments to build the atomic bomb and when they subsequently helped build the atomic bomb. It seems that beginning with those developments, the excitement for science has been allowed to cloud the scientist's decisions to work in areas that adversely affected humanity, and the challenge or emotion has been permitted to bias the judgement expressed by a scientist's authority.

- *Their loyalty to science took a second seat when it was in conflict with their loyalty to their employer.* Such conflicts are inevitable when scientists work for the military, or are under totalitarian regimes, or serve their governments and its institutions. But they are also manifested in the ranks of those who work in industry and are thus restricted by its objectives. There are genuine problems of conflict of loyalty in contemporary industrialized science. Such conflicts of loyalty are murky whether in the chemical, the tobacco, or the electronics industries. They often involve unacceptable patterns of scientific behavior and scientific evidence which is inconclusive. Inevitably they are controversial and political and carry with them an intense scientific conscience.

- *They spoke as knowledgeable experts in areas which transcended their competence.* This is one of the most dangerous and most widely-spread mistakes committed by scientists. They are perceived by society as persons who can speak expertly on virtually any scientific subject and when they are called upon to express judgements on issues that lie outside their domain, they often exceeded the areas of their competence and expressed social judgements using their authority as scientists. In issue after issue, they found themselves engulfed in controversy, whether the issue was the antiballistic missile (ABM) program, the supersonic transport (SST), the effects of nuclear fallout, the merits of and the extent of hazards from nuclear power plants, the effects of man-made chemicals on the ozone layer or global warming, the impacts of DDT and other halogenated hydrocarbons on the environment, or the defoliants and herbicides in the Vietnam War. In these and similar instances where the issues are more social than scientific, scientists often exaggerated to make their case. "If two protagonists claim to know as scientist, through the merits of the methods of science, the one that nuclear testing is essential to the national interest, the other that it is destructive to the national interest, where lies the truth?" pointedly asked B. Commoner.[53] Part of the problem lies in the fact that most such questions contain non-scientific elements and the questions as such transcend science. They can be described and they can be defined in scientific terms but their solutions -- at the time they are posed -- are beyond the bounds of science (see Chapter 10). The scientist thus must guard against allowing himself to be considered an expert when he is not. The public confidence in the objectivity of science and its integrity rests as much on his shoulders as does the advancement of science itself. The various aspects of this issue have been much debated and the two aspects of modern scientist's responsibility (to science and to society) have long been stressed.[54] In these matters, too, scientists are, and behave like, people, no more, no less.

- *Knowingly or unknowingly upheld false* premises. The scientists allowed the prestige of science to be used for questionable purposes and they exercised undue political influence. At times the scientists contributed to excessive expectations of science. For instance, when they oversold promising new weapons to the government -- the X-ray laser for star wars (SDI = strategic defense initiative) being the latest in this series[55] -- or when they failed to deny premises made by politicians on their behalf such as when President Nixon declared his war on cancer. In their multiple recent functions the scientists are involved in politics, in government, in peacemaking, in diplomacy, in international committees, and in multinational enterprises. In such roles they face conflict between the national interest as it is represented by political pressure and their loyalty to the international thrust of science. They are asked to think about fail-safe systems, about weapons inspections, about peacekeeping issues, and to spy on their colleagues. They are exploited by their governments for political advantage and in this they frequently mislead the public through incomplete statements.[56]

- *Scientists short circuited the norms of science.* The most celebrated recent example of this type of regrettable actions by scientists and science workers is the scientific fiasco of cold fusion.[57,58] On March 23, 1989, two chemists -- Stanley Pons of the University of Utah and Martin Fleischmann of the University of Southampton, England -- announced to the press -- not to the scientific community through science's forums and peer review system of publishing -- that they had produced a sustained fusion reaction at room temperature. Their equipment was amazingly simple: a small cell containing a palladium cathode immersed in heavy water at *room temperature.* They reported that in such a cell, deuterium nuclei were fusing and were producing heat at a rate four times higher than the input power. The claim: fusion has been harnessed "in a test tube of water at room temperature." The claim was ultimately shown to be wrong. While the eventual resolution of this incident is a triumph of the method of science and its ability to correct itself, the saga of cold fusion drew attention to an old problem and also to developing new difficulties regarding the norms of science.

The old problem concerns what has often been referred to as *pathological science* ("science" that is not there).[58,59] That is, unexpected "discoveries" and new phenomena not backed by facts, which are prematurely announced. Similar earlier, but less spectacular, examples are the erroneous discoveries of N-rays, polywater, Lysenko's inheritance of acquired characteristics, and Laetrile as a cure for cancer. Sadly, in the case of cold fusion two other of science's strongest norms have been bridged. The first one has to do with the way the discovery was announced and the second one has to do with the way scientists responded to the announced discovery. The two scientists "published" their discovery by press conference (rather than by peer review), by carefully controlling scientific information of premature nature, and by withholding (rather than fully disclosing) of their scientific data. Of equal significance is the fact that in the responses of many notable scientists was absent the characteristic conservative approach and scrutiny normally accorded to important scientific claims. Notable experimentalists and theorists reported confirmation of cold fusion only to retract their claim soon after, in a frenzy to jump on the bandwagon of cold fusion. And the concern stretches even further: scientific isolation in research, unorthodox mechanisms of science funding, and the taking over of the scientific process by profiteers, politicians, administrators, lawyers, and industrial and military opportunists. Circumventing the peer review system of science is full of perils. And yet, other recent scientific announcements were made this way, among them the cloning of sheep and the claim of "life on Mars."

The issue discussed in this paragraph is relevant to, but different from that, of fraud in science. While all forms of breaking down the norms of science are unacceptable, fraud in scientific research is not just unacceptable, it is anathema. And yet, it possibly is inevitable. We referred to a few instances of science fraud earlier in this chapter. Fortunately, the power of science to detect fraudulent results and the ability of science to quickly expose and dispose the guilty is a strong deterrent especially in those cases where the norms of science are feeble or absent.

Facing the New Horizons of the Endless Frontier. Whatever the new frontier of tomorrow happens to be it would demand more of the scientist than in the past. The world of the future will be more science-based and the science of the future will have progressively bigger social involvement. The scientist and the science worker will thus continue to see major changes in their vocation.

The social component of science will increase and with it the social responsibility of the scientist and the science worker. The world crises experienced since WWII (environmental crisis, energy crisis, world-wide food supply, epidemics, arms proliferation, etc.) will multiply, and so will the need for a sound scientific input for their proper resolution. The thought patterns of science and its intellectual and material accomplishments will be brought closer to the political, legislative, and judicial systems through a multitude of existing and expanding avenues that presently include committees of scientists advising directly or through Councils, Academies, Congressional and Executive Offices. The scientist's and the science worker's integrity will be repeatedly tested as they find themselves immersed in socio-political issues and socio-technological functions. Discriminating between what lies within the domain of science and what lies at its fringes, between what is science and what is conscience will undoubtedly be a difficult task for the boundaries are blurred and the landscape is filled with the mines of the corruptive power of power.

The scientists must be forced "out of their aloofness,"[60] directed Dubois. Rational thought is not an excuse of detachment from calamitous issues. A renewal of the scientist's commitment to the norms and the ethic of science and to his individual and collective responsibility to society is thus in order. Science must be made compatible with freedom in -- and prove itself a source of freedom for -- society. And for this the scientist needs to *broaden the definition of his vocation* to include becoming aware of modern man's condition, learning to deal with social complexity and problems, and bringing science and society closer to one another. He needs to commit himself to work for peace and to assume responsibility for the fruits of his work.

The issues of scientific freedom and scientific responsibility will become more numerous and more complicated. Scientific freedom will not be automatic. It will rather be strongly coupled to scientific responsibility. The ethical problems of genetic engineering and those of war and peace will continue to sharpen the contrast between the scientist's freedom and the scientist's societal responsibility.

6. 4. REFERENCES AND NOTES

1. Max Planck, *Where is Science Going?* W. W. Norton & Company, Inc., Publishers, New York, 1932, p. 214.
2. J. R. Oppenheimer, *The Open Mind,* Simon and Schuster, New York, 1955, p. 100.
3. See, for instance, a discussion in G. Taubes, *Nobel Dreams: Power, Deceit, and the Ultimate Experiment,* Random House, New York, 1986.
4. J. Gleick, *Genius: The Life and Science of Richard Feynman,* Pantheon Books, New York, 1992; F. Dyson, Physics Today, November 1992, p. 89.
5. R. Dubois, *Reason Awake,* Columbia University Press, New York, 1970, p. 36.
6. R. C. Hovis and H. Kragh, Scientific American, May 1993, p. 104 .
7. For instance, Rachel Carson, Barry Commoner, Linus Pauling, and others who dealt with critical science on environmental, chemical, radiation, and war issues.
8. Eugene Wigner was noted for his modesty. For many years, he used to visit Oak Ridge National Laboratory where I worked for 31 years. I had the privilege of meeting him on a number of occasions and he was always warm, interested in you as though you were the only one around, eager to share a story, or a piece of news, or to ask a penetrating question in his customary gentle manner. Once, after his retirement from Princeton, in response to my question as to how he was doing, he excitedly related: "You know, now that I have retired from Princeton, everyone thinks that I have lots of time and they ask me to do things. So, I reached a firm conclusion about retirement: *Retiring is tiring!*" Wigner's rule of retirement!
9. This was said of some great scientists, but the classic example is the behavior of Humphry Davy, Faraday's mentor, who so envied his protegee Faraday that he

opposed his election to fellowship in the Royal Society. (E. Segrè, *From Falling Bodies to Radiowaves,* W. H. Freeman and Company, New York, 1984, p. 141).

10. On August 19, 1997 sad and disturbing news hit the national newspapers in America: "A scientist who won a Nobel Prize studying decease.... admitted yesterday in court that he sexually abused a boy he brought to Maryland from Micronesia," wrote The Washington Post (The Washington Post, February 19, 1997).

11. In the opinion of Ziman (J. Ziman, *The Force of Knowledge,* Cambridge University Press, Cambridge, 1976, p. 122), "a highly developed system of prizes, rewards, and public honors for scientific work is dangerous and dishonest. To make 'stars' of a few, and thus, by implication, to degrade those who work hard with less good fortune is not healthy."

12. For example, Segrè (E. Segrè, *From X-Rays to Quarks,* W. H. Freeman and Company, New York, 1980, p. 97) writes about Einstein: "He was not averse to playing the role of the great scientist; clearly he enjoyed it. Perhaps this explains some of his affectations, his strange manner of dress, and some habits that may have been for show. After all, he was an admirer and friend of Charlie Chaplin."

13. It was said of Michael Faraday that he had been "the most enlightened preacher amongst the humble folk whose faith he followed." (R. J. Seeger, Physics Today, August 1968, p. 30).

14. A. H. Guenther, Private Communication, 1997.

15. Such as the statement by Carl Sagan, "The cosmos is all that is or ever was or ever will be," (C. Sagan, *Cosmos,* Random House, New York, 1980, p. 4), or his statements in his preface to Stephen Hawking's book (S. W. Hawking, *A Brief History of Time,* Bantam Books, Toronto, 1988, Preface, p. x) that "there is no absolute beginning of reality and therefore no need for a Creator." Clearly, individual opinions and scientifically unproven assertions presented as, or implied to be, scientific facts.

16. See for example, B. Russell, in *Science, Technology, and Society,* R. Chalk (Ed.), the American Association for the Advancement of Science, Washington, D. C., 1988, p. 4.

17. S. Quinn, *Marie Curie: A Life,* Simon & Schuster, New York, 1995; see also, Physics Today, August 1995, p. 55.

18. A. Pais, *Inward Bound,* Clarendon Press, Oxford, 1986, p. 234-237.

19. E. Teller, Science **121**, 25 February 1955, p. 267.

20. V. Weisskopf, *The Joy of Insight,* Basic Books, New York, 1991, p. 128.

21. V. Ya. Fainberg, Physics Today, August 1990, p. 41.

22. As we mentioned in Chapter 2, A. Einstein urged the United States to construct the atomic bomb. Russian scientists in 1942 also urged their politicians to develop the atomic bomb just like their American counterparts (A. Pais, Physics Today, August 1990, p. 13). The question of whether German scientists during WWII were willing to make atomic bombs for or they conspired to deny Hitler the atomic bomb is still debated (D. C. Cassidy, *Uncertainty: The Life and Science of Werner Heisenberg,* W. H. Freeman and Company, New York, 1992; M. Dresden, Physics Today, June 1992, p. 79). According to Walker (M. Walker, Physics Today, January 1990, p. 52; May 1991, p. 13), "W. Heisenberg flatly stated, he would have considered it a crime to make atomic bombs for Hitler. But Heisenberg also considered it unfortunate that these weapons were given (by the scientists of other countries) to other rulers and were used by them."

23. V. Weisskopf, *The Joy of Insight,* Basic Books, New York, 1991, Chapter 8.

24. Scientific Research 1, December 1966, p. 34.
25. L. Graham, The Sciences, October 1980, p.14.
26. E. Rabinowitch, Impact, Vol. XVII, 1967, p. 107.
27. M. Perutz, *Is Science Necessary?*, Oxford University Press, Oxford, 1991, p. 97.
28. J. R. Oppenheimer, *The Open Mind,* Simon & Schuster, New York, 1955, p. 88.
29. V. Weisskopf, *The Joy of Insight,* Basic Books, New York, 1991, p. 164.
30. M. Perutz, *Is Science Necessary,* Oxford University Press, Oxford, 1991, p. 183; J. D. Watson, *The Double Helix,* Penguin Books, New York, 1968.
31. J. D. Watson, *The Double Helix,* Penguin Books, New York, 1968, pp. 101-104.
32. H. H. Seliger, Physics Today, November 1995, p. 25.
33. M. Perutz, *Is Science Necessary?*, Oxford University Press, Oxford, 1991, pp. 149-163.
34. Great scientists such as Newton, Lavoisier, Helmholtz, Boltzmann, Carnot, Maxwell, Planck, Pasteur, Mendel, Darwin, Arrhenius, and Alfvén met stiff resistance in having their findings and ideas accepted.
35. Pointedly, H. Alfvén described the feud between S. Chapman and K. Birkeland on the theory of magnetic storms this way (H. Alfvén, as quoted by K. Rypdal and T. Brundtland, Journal de Physique IV, France 7, 1997, p. C4-115):

> "Since Chapman considered his theory of magnetic storms and aurora to be one of his most important achievements, he was anxious to suppress any knowledge of Birkeland's theory. Being a respected member of the proud English tradition in science and attending - if not organizing - all important conferences in this field, it was easy for Chapman to do so. The conferences soon became ritualized. They were opened by Chapman presenting his theory of magnetic storms, followed by long lectures by his close associates who confirmed what he had said. If finally there happened to be some time left for discussion, objections were either not answered or dismissed by a reference to an article by Chapman. To mention Birkeland was like swearing in the church."

36. An example of this is the prolonged and ugly confrontation between Samuel Goudsmit and Werner Heisenberg regarding the controversy over the German atomic bomb (M. Walker, Physics Today, January 1990, p. 52; Physics Today, May 1991, p. 13).
37. C. P. Snow, in *Science, Technology, and Society,* R. Chalk (Ed.), the American Association for the Advancement of Science, Washington, D. C., 1988, p. 6.
38. Characteristically, Wilson (M. Wilson, The Atlantic, September 1970, p. 101) writes:

> "Years ago, as a graduate student, I was present at a three-way argument between Rabi, Szilard, and Fermi. Szilard took a position and mathematically stated it on the blackboard. Rabi disagreed and rearranged the equations to the form he would accept. All the while Fermi was shaking his head. 'You're both wrong,' he said. They demanded proof. Smiling a little, he shrugged his shoulders as if proof weren't needed. 'My intuition tells me so,' he said. I had never heard a scientist refer to his intuition and I expected Rabi and Szilard to laugh. They didn't. The man of science, I soon found, works with the procedures of logic so much more than anyone else that he, more than

anyone else, is aware of logic's limitations. Beyond logic there is intuition .... "

Many other examples can be cited. For instance, Boltzmann's physical intuition was legendary, and P. Dirac, a worshiper of mathematical logic, was a master of intuition.[6]

39. Rutherford is a classic example of a scientist who was sensitive to small anomalies. He pursued them relentlessly, often to revolutionary conclusions. This was the case of the nitrogen anomaly which led him to the nuclear transformations, and the case of the tiny proportions of alpha particles which were scattered at large angles that led him to the discovery of the nucleus.

40. J. T. Edsall, Science **188**, 16 May 1975, p. 687.

41. Science **184**,10 May 1974, p. 644.

42. Science **212**, 10 April 1981, p. 137.

43. W. Broad and N. Wade, *Betrayers of the Truth,* Simon and Schuster, New York, 1982.

44. S. G. Brush, Science **183**, 22 March, 1974, p.1164.

45. R. J. Seeger, Physics Today, August 1968, p. 30.

46. In 1990, I was attending a meeting at the United States National Academy's complex in Woods Hole, Massachusetts. In one of the sessions I was sitting by Professor Michael Kasha of Florida State University listening to a lecture by a distinguished colleague. After listening to the talk for sometime, I turned to Professor Kasha and asked: "Do you understand what he has been talking about?" Kasha replied: "You know, 20 years ago Robert Platzman was sitting next to me listening to the same speaker and he asked me the same question!"

47. C. P. Haskins, American Scientist, **58**, January/February 1970, p. 23.

48. T. S. Kuhn, *The Structure of the Scientific Revolutions,* The University of Chicago Press, Chicago, Illinois, 1970.

49. M. Polanyi, in *Physical Science and Human Values,* Princeton University Press, 1947, p. 124.

50. M. Polani, Minerva V, Summer 1967, p. 533.

51. L. Kerwin, Science **213**, 4 September 1981, p. 1069.

52. A. Szanton, *The Recollections of Eugene P. Wigner as Told to Andrew Szanton,* Plenum Press, New York, 1992, p. 313.

53. B. Commoner, *Science and Survival,* The Viking Press,, Inc., New York, 1967, p. 106.

54. See for example, A. Weinberg, Minerva **16**, 1978, p.1, and J. T. Edsall, Science **212**, 3 April 1981, p. 11.

55. D. Blum, Bulletin of the Atomic Scientists, July/August, 1988, p. 7.

56. F. von Hippel and J. Primack, Science **177**, September 29, 1972, p. 1166.

57. F. Close, *Too Hot to Handle, the Race for Cold Fusion,* Princeton University Press, Princeton, New Jersey, 1991.

58. J. R. Huizenga, *Cold Fusion: The Scientific Fiasco of the Century,* University of Rochester Press, Rochester, New York, 1992.

59. I. Langmuir (Transcribed and edited by R. N. Hall), Physics Today, October 1989, p. 36.

60. R. Dubois, *Reason Awake,* Columbia University Press, New York, 1970.

CHAPTER 7

From Basic Research to Application (Science and Technology)

Although the boundary between knowledge and its use is blurred and the translation of scientific findings to use by society is a complicated and still not a completely understood process, today's technology *is* the product of science and concomitantly today's technology *is* a critical element of the advancement of science. To quote a modern technologist[1]: "Today's technology is based on yesterday's science; today's science is based on today's technology. The science that even now is making discoveries that will create new industries cannot be done without, for example, the lasers and computers that have been developed from previous science." Indeed, for a number of "high-tech" industries, today's technology is based on today's science.

7.1. SCIENTIFIC RESEARCH AND TECHNOLOGY

Prior to elaborating on the intricate ways science and technology interact, let us expand a little on the terms used to describe various types of scientific research and on technology itself.

7.1.1. Types of Scientific Research

Scientific research is broad. Depending on its goal it has been described by different names: basic research, applied research, mission-oriented research, problem-oriented research, industrial research, etc. Some[1] simply divide scientific research into two categories: fundamental (basic and applied) and strategic (basic and applied, but aiming at a particular applied goal). The basic-applied-development spectrum has no distinct boundaries and normally stretches considerably over time.

Basic Research. The term "basic" or "pure" research is used to describe scientific research that is performed without thought of practical ends. This type of research is not easily defined operationally and cannot be tested in advance for utility. The process of innovation is interwoven with the production of new knowledge. Indeed, it has been said that basic research is the mother of all inventions because it provides the scientific capital (new scientific knowledge and understanding) that is needed for technological breakthroughs and the finding of solutions to important practical problems. In the modern world, basic research is the "pace-maker of technological progress." Technology upon technology originates from fundamental discovery (see Section 7.2) often in ways unforeseen. Rutherford, as we noted earlier, considered the energy produced by the breaking of the atom "a very poor kind of thing,"[2] and Einstein did not foresee how his mass-energy relationship would lead to the release of nuclear energy. In contrast, Faraday foresaw the practical use of his work on electricity and magnetism. When (around 1850) the then-chancellor of the exchequer William Gladstone after visiting Faraday's laboratory asked "This is all very interesting, but what good is it?" Faraday's reply was visionary: "Sir, I do not know, but some day you will tax it." Indeed, electrical technology was the first important science-based activity to be developed from the beginning on scientific principles. Science came first and owed virtually nothing to social needs or practical techniques (in contrast to the steam engine[3]).

It is argued that industrial development would eventually stagnate if basic scientific research were long neglected, and that the consequences of neglecting basic research are normally felt when the scientific capital runs out. Clearly, not everything that can be known about a given subject must be known to be able to do things (we would not have gone to the moon if we waited to learn all that has to or can be known about the subject), but to find solutions to certain problems -- e.g., to find a cure for cancer or an aids vaccine -- requires first broad basic research. Solutions are preceded by knowledge, not by chance, decree, or legislation. In fact, basic research is the foundation of modern technology for a number of other reasons besides the basic knowledge it provides. It is the foundation of education and the basis of training those working in industry and technology. It cultivates a scientific climate conducive to understanding of and critical and objective about technology.[4]It is a source of intellectual standards for applied research. Basic research is the net exporter of technique to industry. Techniques now common in industry, such as vacuum technology, cryogenics, X-ray diffraction, radioisotopes, and nuclear reactor and neutron spectroscopy instrumentation had their origin as techniques of basic research. Basic research, then, must not be a peripheral activity or be forced to provide short-term solutions under excessive pressure and limited support.

Applied Research. Applied scientific research is done to achieve certain ends, to translate the findings of basic scientific research into practice by providing "complete" answers regarding the applied end. Applied research precedes engineering and industrial development. It is best carried out when it is in contact with basic research and when the applied scientist is both knowledgeable in fundamental science and skilled in application and is not narrowly channeled. Moreover, since most of technology can be traced to Little Science, applied science must be effectively coupled to Little Science. An innovative society therefore must think big about supporting basic Little Science.

Applied science leads to inventions. The latter stretch out over time and can involve many individuals. The idea of generating coherent radiation at optical frequencies for instance was conceived in late 1957 and by the end of 1960 there were five realizations of the laser.[5] Conversely, applied research leads to basic. Radar and radio astronomy would probably have never been undertaken for their own sake if large radar antennas had not been built originally for applied purposes. Similarly, development of pure materials for technological applications has stimulated fundamental investigations in solid state.

Mission-Oriented and Problem-Oriented Research. Mission-oriented research is generally broad research in support of a mission, a technological goal such as, for instance, the development of radar, nuclear energy, controlled thermonuclear reactions, X-ray lasers, understanding the effects of radiation on matter, space exploration, finding a cure for cancer, and so on. Basic research benefits greatly from such well-thought-out mission-oriented endeavors. For instance, basic research in superconductivity draws support from missions dealing with the development of new sources of energy; atomic and molecular physics are supported by governments for their relevance in the space program; atomic, molecular, radiation, radiological physics and chemistry draw support from agencies whose purpose is the understanding of the effects of radiation on living matter. It is also clear that the specialized interests of mission-oriented agencies stimulate new instrumentation and new experimental techniques which might not have otherwise been generated. For example, the scale of support for computers for defense and space purposes benefited all science. This benefit might not have been realized if computers had been developed exclusively for their scientific value. Even missions that fail are sometimes successful! Such, for instance, might be considered President R. Nixon's war on cancer initiative which did not provide the cure for cancer it was intended for, but which set in motion the new and important advances in biotechnology. A similar example might be considered President R. Reagan's star wars initiative which did not deliver the protective shield against nuclear attack it sought, but which led to research efforts for new materials and a push for new light sources such as X-ray lasers. The mutual benefit has been the most when the interpretation of mission relatedness of basic research was not narrowly defined.

Problem-oriented research is narrow research specifically directed to the solution of a particular problem or need. It may seek quick solutions (technological fixes[6]) to pressing societal needs. Problems, for instance, relating to health protection, biomedical analysis applications, pollution, replacement of useful but hazardous materials [e.g., PCBs (polychlorinated biphenyls), CFC's (chlorofluorocarbons)], utilization and conservation of energy, transportation, water supply, waste disposal, and so on.

Industrial Research. Products are produced by industry, not by science. Science enables industry to develop new technologies and to reduce scientific discovery to practical applications effectively and quickly, but it is industry that produces products. Industry itself needs research that predates invention, and people who are challenged by devising ingenious ways that facilitate the emergence of new technology from their scientific knowledge and insight. Industry, therefore, undertakes research of its own. This is referred to as industrial research. Getting good new ideas is the ultimate objective of industrial research. Therefore, research is performed by industry because new knowledge can be the seed for a good idea or the basis for converting an idea into a good one, and thus to increased future profits. There is a time continuum from fundamental knowledge to an actual product which is getting shorter as the few examples we presented in Table 3.12 of Chapter 3 would attest. And there is a time period for the diffusion of technological innovations which is also getting shorter with time. For instance, the time required for the *diffusion* of the steam engine was 150 to 200 years, for the automobile 40 to 50 years, for the vacuum tube 25 to 30 years, and for the transistor about 15 years.[7]

Fundamental and Strategic. Following Richter[1] one may simply divide scientific research into two broad generic areas: fundamental and strategic. Both have basic and applied components. Richter uses the laser as an example to make the distinction between these four categories of research (see Table 7.1).

Table 7.1. Examples of generic, basic, and applied research for the case of the laser[1]

Generic research	Types of research	Examples of research for the case of lasers
Fundamental		
	Basic	Quantum mechanics (the Einstein A and B coefficients for light absorption and emission)
	Applied	Laser
Strategic		
	Basic	Interaction of materials with light
	Applied	Optical fibers

7.1.2. Technology

Technology is the organization of knowledge for practical purposes. It meets man's need to do. It embraces the knowledge, the methods, the techniques, and the devices man uses to exploit the environment to his benefit and to satisfy his needs. Through technology man goes beyond what nature herself provides for him and manufactures his own materials as his needs demand. The term technology originates from the Greek word *techne* (τέχνη), meaning art, skill, a practical method of making things.

Technology is much older than science. Man did not always have science, but he did always have some form of technology. Indeed, there has been -- and there still is -- technology which is not science based such as in the arts (e.g., sculpture and music) and the crafts (e.g., cooking and tailoring).[8] There is, as well, a great deal of nonverbal thought in technology. "Pyramids, cathedrals, and rockets exist not because of geometry, theory of structures, or thermodynamics, but because they are first a picture in the minds of those who built them," it has been said.[9] *Homo sapiens* (man the thinker) was first *homo faber* (man the maker).[10] Man made tools and tools helped make him. He made tools to handle and work with fire, agriculture, and metals, and "machines" such as the plow 7,000 years ago,[11] and wheel vehicles some 5,000 years ago.[12] None of these -- not even the steam engine[13] -- had the help of scientific knowledge. But this is no more. The very terms by which the progress of civilization was measured (stone age, bronze age, iron age) that were technology-based have now been replaced by science-based terms: the atomic age, the nuclear age, the computer age, the information age, and the biological age. Just the same, technology-based and science-based terms have a distinct commonality: the development of a technological mastery by man of his environment.

Since the 17th century man learned to systematically use scientific knowledge for doing. He established technologies as systematic disciplines to be taught and to be learned. In due course he reorientated science toward breeding these new disciplines of technological applications. As a result, the industrial revolution changed man's ways of working and living, revolutionized virtually every aspect of his life, changed the tempo of his existence, and paved the way for even further and faster changes in the 20th century. Modern science-based technology[14] besides the jet engine, the computer, the cellular phone, and the modular product it provided, it is called upon to solve virtually every problem of modern society, from war, to poverty, to population and disease, to economic development. Through science, technology has made human existence infinitely more secure in terms of health, epidemics,[15] and even in terms of natural catastrophes, and yet simultaneously it made human existence infinitely more dangerous in terms of human conflict and the consequences of man's irresponsibility and carelessness. It impacted man's values, political systems, and human conditions. It made democracy more representative and brought regional social inequities to universal attention. It is facilitating the unification of the peoples of the earth in much the same way in the late 19th and early 20th centuries it facilitated along with industrial expansion the curses of communism and imperialism. Most significantly, *science-based technology has made the superpower's panoply too heavy to use with impunity and the high-tech communication the people's ultimate means of civil disobedience.*

7.1.3. Similarities Between Science and Technology

Like science, technology is a human activity. Like science, technology too is blamed for things that belong to conscience. The character of both is revolutionary and complex. It is in the nature of both to solve human problems and to generate new ones. They both open up possibilities for society which had not been there before. Some of these new possibilities society loves (e.g., TV), others it needs (e.g., synthetic materials), yet others (e.g., nuclear power) it needs but does not want. The conditions for technology have changed just as they did for science. For instance, the further development and future use of nuclear technology are not determined by what science-based technology can or cannot do, but rather by society's attitude toward it, an attitude reflective in large measure of society's lack of confidence in itself. Nuclear technology is limited by society's inability "to exert the eternal vigilance needed to ensure proper and safe operation of its nuclear energy system," wrote Weinberg.[16] A new order indeed! New conditions and new boundaries which are progressively spreading to other sectors of technology, including the electrical, chemical, and biomedical industries.

In spite of the momentous recent advances of science-based technology, man's tailoring of the world to his liking through technology is as limited as is his understanding of physical reality through science. Like science, technology is limited[17] by the laws of nature (e.g., by the universal conservation laws), is constrained by the state-of-the-art (e.g., deriving electric power from nuclear fusion sources would violate no law of nature, but it is presently beyond the state-of-the-art), and is restricted by the structure of society and the political and legal systems.[18] Like science, technology is handicapped by the errors of its practitioners[19] and by the curses that accompany its blessings.

7.1.4. Differences Between Science and Technology

There are differences in the method, goals, and operation of science and technology, as well as between the scientist and the engineer. It is instructive to ponder over a few of these differences.

- The functions of science and technology are different. Science works autonomously and almost independently of society. Although technology also has a way of its own, how technology works is largely determined by the standards set up by society. The demands for technology are pressed from below (extrinsically). In science the demands are largely set from within (intrinsically).

- Their *modus operandi* is different. Rarely in science things are done twice in the same way. Industry does exactly that. The blue print of industry is the "blue print."

- Their environments are different. Science needs an intellectual setting which is mostly determined by the institution in which it is situated. While this may be true to some extent for technology as well, for technology to be useful it has to be socially situated.

- Their constituencies are different. The constituency of science is society at large. The constituency of technology is the customer. Technological decisions are based on markets, on profit considerations, on what the consumer wants or on what the consumer can be manipulated to want. Innovation would not be accepted if there were no market for it.

- Their goods are different. The end product of science is knowledge, the scientific paper and the accumulation of literature. The end product of technology is a machine, a chemical, a process. The product of science is public good par excellence.[20] As such, it is provided on an all-none basis and consumed in a joint way: more for one consumer is not less for another. In principle, all humanity shares in the findings of basic scientific discovery at no extra cost (even a meteorological forecast, once made, benefits everyone at no additional cost, noted Mesthene[21]). The product of technology is different: it normally is a consumer product of which every additional unit adds to the cost of production. In contrast with the public goods of science, the goods of technology are private, consumer based.

- Their cultures are different. Within the science culture, the results of the scientific research are considered free goods. While a scientist's research may be kept secret during the course of the investigation and scientists compete in their search for discovery, once a discovery has been made, revealed, and confirmed, it belongs to humanity. In contrast, technology is developed in secret. Publication is incidental and it is often replaced by the patent. The scientist is anxious to tell the world of his findings with every conceivable means of communication, the technologist works hard to prevent the spread of his new knowledge. Thus, the final test of validity in technology is public use and the material advancement of society. The drive for secrecy in technology is so strong wrote Baruch,[22] "that early public policy created the U.S. patent system. Society went so far as to grant a monopoly to the technologist in exchange for revealing the technical knowledge embodied in his patent's disclosure."

- Their support is different. Technology pays for itself in a direct way. Science pays for itself in an indirect way. The engineer's livelihood results directly form the product of his work. The scientist's livelihood results from the presumed value of his work to society and/or the contribution his work makes to the underpinning of technology in its pre-competitive stage. The products of science-based technology more than pay for the cost of scientific research. As Funder wrote[20] in 1979, "the discovery of effective polio vaccines alone saves the US community more each year than the entire medical research budget." Indeed, how could one put a price on Louis Pasteur's discovery of the germ theory (1860), possibly the greatest single advance ever made in medicine?

- Their practitioners are different. Naturally, science has more scientists and technology has more science workers. They both need freedom but the former need more and broader freedom than the latter. Even in industrial research success comes through good individuals who are left alone. As we stated earlier, the technologist and the engineer are motivated externally by society while the scientist's motivation is internal. Thomas A. Edison on October 21, 1879, produced the first practical electric light, perhaps the most significant of his many inventions, and Enrico Fermi on December 2, 1942 built the first nuclear reactor, perhaps the most extraordinary product of his long career in basic research. There is a difference between the two men and their craft, but there is also a fundamental similarity: they and their work changed man's life profoundly and irrevocably.

7.1.5. The Debt of Science to Technology

The relationship between science and technology has been described by some as incestuous. Technology is viewed as both the mother and the daughter of science. Independently of the proper description of their relationship, there is a mutual debt and feedback between the two that grows with time. "The road from science to new technologies is not a straight highway but a kind of spiral of science enabling new technologies that, in turn, allow new science that again creates new technologies and so forth."[1]

The debt of science to technology is many-sided. It embraces contributions of technology to individual fields of science, generic transdisciplinary impact across much of physical science, and ideas generated in technology that cross over to science.

In every step in the *step-by-step* process of science one finds a host of particular technologies that enabled science to make the next step. For instance, critical steps forward in astronomy, physics, chemistry, and biology more often than not had as a prerequisite the previous existence of required minimum technology: in astronomy, successive generations of telescopes, satellites, remote sensing devices; in physics, spectrographs, accelerators, nuclear reactors, exotic materials; in chemistry, new chemicals, analytic instruments, spectrometers; in biology, radioactive tracers, absorption, fluorescence and scintillation spectrographs.

The technologies of vacuum, light, and the electron have particularly impacted science in a generic transdisciplinary way.

Vacuum Technology. Scientists, especially physicists, love empty space. The reason for this is simple: vacuum is the ultimate environment to study reacting species in isolation. It is necessary to the reductionist approach of basic science. The pumping-down of tubes or chambers to subatmospheric pressures was a necessary tool in the early science of gas discharges, themselves having been the precursors of the discoveries of the electron and X-rays. The needs of the electric light industry and the manufacturers of radio tubes among others have contributed to the development of high-vacuum technology which is now one of the most significant requirements for many advanced fields in science for the study of particles, gases, surfaces, and plasmas. Vacuum technology itself has had a miraculous success. From vacuum levels of 10^{-3} Torr in the 1900's, it reached vacuum levels of 10^{-11} Torr in the 1970's, and levels of 10^{-16} Torr today. Without the technology of vacuum a great deal of science would not be possible.

Light-Source Technology. Much science owes its advancement to light-source technology. This technology is especially science-based. It moved progressively from primitive light sources to advanced light sources of varied intensities, durations, and spectral compositions, and to the laser and the synchrotron light. The most advanced science (basic and applied) today whether in physics, chemistry, or biology is benefitting from these developments. Light-source technology parallels that of vacuum technology in its success with regard to the duration of light pulses. Over the last forty years the duration of light pulses has decreased from milliseconds (10^{-3} s) to femtoseconds (10^{-15} s).

Electrical and Electronics Technologies. Without electrical technology, science -- especially physics -- would not have started and would not have advanced the way it did. How could one envision the major discoveries in physics without the high voltage power supply, the current supply, the electrical instruments, the power conditioning instruments developed for technology and obtainable cheaply because industry needs them in large numbers? The complicated electronic devices which crowd every physics research laboratory for instance would be impossible without the cheap components manufactured originally for radio, TV, and computers.

Computers and Data Processing Technologies. This technology made possible the recent capabilities in scientific data acquisition, processing, and storage. It facilitated the effective communication in science, and allowed extraordinary information dissemination and compression. It enables the scientist to cope in some way with the proliferation of his semantic environment.

Finally, technology impacted science through ideas. As an example, we are reminded of the industrial scientists who were awarded the Nobel prize in physics or chemistry for work they did in industrial laboratories[23] (see Table 7.2).

Table 7.2. Five Nobel prizes awarded to industrial scientists

Industrial scientist(s)	Discovery/Contribution	Industrial laboratory	Year
I. Langmuir (Chemistry)	Surface chemistry/ Electrical discharges/ Atmospheric physics	General Electric Research Laboratory, Schenectady, New York	1932
J. Bardeen[a]/ W. H. Brattain/ W. Shockley (Physics)	Transistor	Bell Telephone Laboratories, New Jersey	1956
A. A. Penzias/ R. W. Wilson/ (Physics)	Cosmic background radiation	Bell Telephone Laboratories, New Jersey	1978
G. Binnig[b]/ H. Rohrer (Physics)	Scanning tunneling microscopy	IBM,[c] Zurich Research Laboratory	1986
K. A. Müller/ J. G. Bednorz (Physics)	High temperature superconductivity	IBM, Zurich Research Laboratory	1987

[a] Bardeen also shared the 1972 Nobel prize in physics for his work on the theory of superconductivity.

[b] The third scientist who shared the Nobel prize (E. Ruska) was not at IBM.

[c] International Business Machines.

7.2. THE TECHNOLOGICAL VALUE OF BASIC RESEARCH

As we have mentioned earlier (see Table 3.9), the first half of the 20th century is the time period in man's evolution during which successive fundamental discoveries in science were followed by technological breakthroughs and industrial innovations. Even remote areas of physics, such as the cyclotron, spark and bubble chambers, and experimental advances related to elementary particles, had led to or accelerated practical innovations in magnetism, circuitry, pulsed energy sources, new materials, cryogenics, superconductivity, and so on. From basic scientific knowledge before WWII came space vehicles, nuclear explosives and power plants, new substances, and electronic machines. Since then along with the microminiaturization of electronic devices and food packaging from space exploration, came "synthetic" atoms (plutonium, strontium), "synthetic" molecules (freons, DDT), and "synthetic" microorganisms (recombinant DNA). Since then one after another the instruments of technology have been extended through the scientist's attempts to sharpen the tools of his science for experimentation at the frontiers of science.

7.2.1. Basic Discoveries that Changed the World

The link between science and technology can be further illustrated by a number of scientific discoveries that changed the world. Examples of science-based technologies that trace to such discoveries in the fields of electricity and electronics, energy, radiation, chemistry, biomedicine, laser and photonics, and materials are briefly described below.

Electrical and Electronic Technologies. About thirty five years after Faraday's basic scientific discovery of electromagnetic induction (1831) came the development of the first commercial electric generator (1866-67). With that development, electricity during the latter part of the 19th century was transforming not merely the study of physics but also European and American society. In an unprecedented way electricity bridged the gap between pure science and useful applications. It showed the utilitarian character of physics just as chemistry's utility had already been demonstrated in agriculture and industry. Hence, electrical engineering emerged as the first important activity to be developed from the beginning on scientific principles. Since then the science of electricity has given society electric discharge tubes, electric lights, electric motors, telephones, radios, televisions, and clean, reliable, technology-tailored electric power without which there would be no computers and no communication-systems industry as we now know them. The electronics industry came after the discovery of the electron, the induction coils in motor cars came after the laws of induction, the electromagnetic waves and communications came after their discovery by Maxwell and Hertz, and the transistor came after the basic research in condensed matter and quantum theory of solids. Similarly, basic circuits in computers originated in nuclear physics research in the 1930's by scientists who needed to count nuclear particles. The impact of scientific discovery in this field on advanced technology continues with the miniaturization of electronic devices and computer microprocessing.

Energy Technologies. As was discussed in Chapter 2, man's most important energy sources are science-based. They will become more so in the future. Nuclear power came after and not before nuclear physics. Energy from controlled fusion is not yet available to man because basic science -- say, in plasma physics -- is not yet sufficient to allow technology to proceed. It was not technology but basic science that formulated the understanding and identified the critical reactions in both fission and fusion which man can harvest for useful energy production. Plasma physics is central to thermonuclear research and to the applied science which is needed to enlarge mankind's energy resources.

Energy technologies have many aspects. They not only involve energy production, but also energy use, energy conservation, energy conditioning, and energy transmission and distribution. Especially in the last area there is a great potential for *superconductivity*. The power loss in a superconducting transmission line would be virtually zero because the electrical resistance of a superconductor is virtually zero. Power transmission by superconductors will become commercially attractive when the savings on power loss exceed the cost of refrigeration of the line, *but also* when room temperature superconductors are discovered and developed. The continuing development of superconducting alloys with higher working temperatures provides hope that the economic crossover may soon occur thus allowing economic long-line transmission of power from distant hydroelectric or other-type plants. Another important application of superconductivity is in very-high-field electromagnets. Such magnets are needed in plasma containment, a key element in the development of a controlled thermonuclear reactor. High-temperature superconductor-based technologies are not here yet because basic science has not yet developed an understanding of the phenomenon that would allow applied research to provide the complete answers needed for their technological development. This technology too will follow and will not precede science. Industry or society did not dream superconductors. Science discovered the phenomenon, struggles to understand it thoroughly, and when it does application will follow and so will the superconducting transmission line and the high-field electromagnets.

Radiation-Based Technologies. Here again we have beautiful examples of scientific discoveries that led to new technologies.

- X-rays. Roentgen's researches on electrical discharges in gases at the end of the 19th century led to the discovery of X-rays and with it to a multitude of technologies in medicine and elsewhere. The latter followed the former.

- Radioactive tracers. They came from nuclear physics and profoundly impacted society via the many technologies in medicine (nuclear medicine for instance) and biochemistry. Many advances in molecular biology would not have been possible without radioactive tracers.

- Radioisotopes. They are widely used throughout science, technology, and medicine. The ability to detect, measure, understand, and safely use ionizing radiation came from science too. It has even revolutionized archaeology by making possible to date more precisely human artifacts and other remains.

- Magnetic resonance imaging. Developed in the 1980's, it came from fundamental work on nuclear magnetic moments in the late 1930's.

Chemistry-Based Technologies. It has correctly been said that of all the branches of science chemistry is the closest to industry. Indeed, the strong coupling of chemical science to technology is responsible for today's chemical environment.

- Chemical synthesis delivers annually about a quarter of a million new compounds, more than 1,000 of which reach the market place.[24] It gave society biodegradable detergents, agricultural, industrial, and medical substances, along with penicillin, vitamins, and hormones. It gave birth to biotechnology and hopefully, by synthesizing the organic and inorganic superconducting materials, to a superconductor industry.

- Chemistry-based technologies handed society plastics, fibers, rubbers, coatings, adhesives, films, and polymers. Out of basic research in theoretical, structural, quantum, and computational chemistry on simple, complex, and polymeric molecules, and through the use of a broad spectrum of experimental techniques, grew the industry of plastics (e.g., Polythene, Lustrex, Lexan, Zytel), artificial fibers (e.g., Nylon, Fortrel, Orlon, Dynel, Rayon, Dacron), and synthetic rubber (e.g., Natsun, Neoprene, Hypalon). The impact upon the standard of living of society of these comparatively inexpensive materials has been immeasurable. And there is more to come. From basic research in chemical dynamics comes understanding of the mechanisms of enzymatic action controlling the chemistry of life, and from quantum and computational chemistry come powerful new tools for pharmacology and emerges an understanding of the interactions with the body of drugs, chemical carcinogens, metals, and other dangerous toxic substances. From basic scientific research emerges a new generation of chemical technology capable of microprobing life at the cellular level.

- Through the data provided by scientific research and the aid of the computer, chemistry has built databases which make possible the *fingerprinting* of complex biostructures. And through extra sensitive analytical instruments rooted in scientific research it is now possible to detect (at the parts per trillion level) and to characterize trace chemicals in diverse environments whether these are environmental pollutants, dangerous biochemicals responsible for rare diseases, or explosives used by terrorists.

Science-Based Biomedical Technologies. Modern health care from immunization, pain killers, chemically-controlled-body changes, to electrical recordings from the brain or the heart, to control of fertility is largely based on past results of fundamental research. Basic science is also responsible for instruments and methods developed by modern technology for use in biomedicine whether these measure electric current, voltage, charge, magnetism, photon fluxes or photon energies, or high-energy radiations. Such are technologies dealing with separations, X-ray, γ-ray, and particle-beam sources, radioactive isotopes, medical scintillation spectrometers, microscopes, cryogenic equipment, fiber-optic sensors a fraction of a millimeter in diameter for *in vivo* measurements,[25] and laser beams to repair detached retinas, seal leaky blood vessels in the eye, treat ulcers, skin tumors or special types of surgery,[26] and microsurgical operations on single cells.[27] The list can stretch further. For instance, with the aid of techniques from physics and chemistry, scientific discovery in biology facilitated new biological engineering technologies and through them new biologically-based therapies such as gene therapy, manipulation of the immune system, and defects in tissues or organs.[28]

Laser-Based Technologies. This is an example of scientific knowledge laying dormant until scientific advances in neighboring areas and technological needs in neighboring fields made its development inevitable. Indeed, the process of stimulated emission of radiation has been shown to be possible in 1917 by Einstein and thus since that time light amplification and the invention of the laser were in principle possible. The laser however was not invented until after WWII when as a result of the development of radar during WWII and the extension of that work to higher microwave frequencies conditions were explored under which laser action can be achieved. Thus, in the early 1950's came the invention of the maser (Microwave Amplification by Stimulated Emission of Radiation) and in the late 1950's the extension of maser principles to the optical region of the electromagnetic spectrum. By 1960, a number of groups were investigating systems that might serve as the basis for the optical maser or laser (Light Amplification by Stimulated Emission of Radiation).[29] Today materials for lasers are many and include gases, liquids, and solids. Lasers come in many varieties, power levels, wavelengths (infrared, visible, ultraviolet, and possibly also X-ray), and types (continuous or pulsed).

Lasers led to new technology which in turn facilitated new science, which again led to new technology and yet again to new science -- a continuous interplay that is still unfolding. High-quality lasers and hardware can now be purchased readily enabling laser-based technology to be used in virtually everything: medicine (see preceding paragraph), industry (e.g., cutting, drilling, welding), communications (e.g., via satellite, fiber optic,[30] or laser printing), weapons (e.g., directed energy weapons), information storage[31] (laser recording, optical disk storage), remote sensing, and so on. Laser-based technologies are used in microstructure engineering, microfabrication, semiconductor processing, material deposition and etching, and a host of methods for altering the morphology of a solid surface with spacial resolution down to the nanometer scale.[32] Very-high power lasers have a potential application in fusion energy sources,[33] and short-duration laser pulses[34] are basic to man's ability to modify and/or switch material properties.

Science-Based Materials Technologies. If we use the term "materials" to refer to solids (naturally occurring or artificially produced) needed by man to manufacture the things he wants, then it can be said that all man needs to do today is to specify the property of the material he needs and science will find the chemical or physical method to get the best improvement in whatever property is sought. Science-based technology gave man the electric light bulb filament, the transistor, the solid-state laser, composites, ceramics, metals, alloys, and polymers.

The discovery of new techniques for producing and processing materials continues unabated and is joined by new capabilities toward the development of new multi-property materials.[35] For instance, materials that show unique physical properties: conductivity, superconductivity, optical effects, magnetism, heat sensitivity, and so forth, or materials that can be made to change their properties, for instance, from insulators to conductors and from conductors to insulators when they are exposed to physical insults such as laser light, or still materials whose three-dimensional structure would allow information processing to occur in bulk rather than surfacially as in the silicon chip.

7.2.2. The First Stretch on the Long Road that Leads from Basic Research to Application

The road that leads from basic research to application can be illustrated by many examples. In Appendix C we opted to describe this by two examples of basic scientific findings in a small field of Little Science, namely, low-energy electron collision physics. These examples involve the development of efficient CO_2 lasers and the development of gaseous dielectric materials for the transmission and distribution of electricity. They and innumerable other examples of the translation of scientific findings to technological products allow us to conclude: *what is good science can be good technology.*

Clearly, the reciprocal feedback between science and technology is overpopulating the earth with offsprings. This process will undoubtedly continue and along with it the shrinking of the time that is required to go from basic research to application. It would appear (see Table 3.12) that this time may be decreasing to virtually zero. Indeed, this may already be happening in the information and computation technology.

7.3. A FINAL THOUGHT

It is hoped that through science-based technology a bridge is constructed to a brighter future that would alleviate the condition of man. Interwoven with this promise are two important elements that need stressing. First, whether science-based technology will at last provide for all of the earth's people and make the earth a more equitable place will depend more on man's values and less on his science and technology. Second, man must become knowledgeable enough, wise enough, brave enough, and unselfish enough to forego technologically brilliant ideas when they are more damaging than beneficial.

7.4. REFERENCES AND NOTES

1. B. Richter, *Sci-Tech Information*, National Institute of Standards and Technology, Vol. 11 (No. 10), October 1995, p. 12; Physics Today, September 1995, p. 43.
2. D. Braben, *To be a Scientist*, Oxford University Press, Oxford, 1994, p. 12.
3. J. Ziman, *The Force of Knowledge*, Cambridge University Press, Cambridge, 1976, p. 31.
4. H. Brooks, Science **174**, 1 October 1971, p. 21.
5. J.Bromberg, Physics Today, October 1988 , p. 26.
6. The *technological fix* concept (A. M. Weinberg, *Reflections on Big Science*, The M.I.T. Press, Cambridge, MA 1967, p. 141)focuses on social problems which are well within the grasp of existing technology without waiting for (or excluding)comprehensive and broader future solution(s).
7. D. A. Schon, *Beyond the Stable State*, Random House, New York, 1971, p. 24.
8. J. K. Feibleman, Nature **209**, 8 January 1966, p. 122.
9. E. S. Ferguson, Science **197**, 26 August 1977, p. 827.
10. M. Kranzberg and C. W. Pursell, Jr., in *Technology in Western Civilization*, M. Kranzberg and C. W. Pursell, Jr. (Eds.), Oxford University Press, New York, Vol. 1, 1967, p. 739.
11. R. Dubois, *A God Within*, Charles Scribner's Sons, New York, 1972, p. 112.
12. R. J. Forbes, in *Technology in Western Civilization*, M. Kranzberg and C. W. Pursell Jr.(Eds.), Oxford University Press, New York, Vol. 1, 1967, p. 34.
13. The primitive steam engine was invented by Newcomen in 1705 and was much improved by James Watt in 1765 (R. Dubois, *Reason Awake*, Columbia University Press, New York,1970, pp. 82 and 83).
14. The United States National Academy of Engineering in December 1989 on the occasion of its 25th anniversary listed (National Academy of Engineering, *Engineering and the Advancement of Human Welfare, Ten Outstanding Accomplishments 1964-1989*, Washington, D. C., December 5, 1989) the following as the ten most outstanding achievements between 1964 and 1989: moon landing, application satellites, microprocessor, computer-aided design and manufacturing, CAT (Computerized Axial

Tomography) scan, advanced composite materials, jumbo jet, lasers, fiber-optic communication, and genetically-engineered products.

15. Artificial immunization has virtually eliminated diphtheria, whooping cough, poliomyelitis and tetanus in the United States, and smallpox worldwide (The National Research Council, *Science and Technology*, W. H. Freeman and Company, San Francisco, 1979, p. 85).

16. A. M. Weinberg, Science **177**, 7 July 1972, p. 27.

17. E. E. David, Jr., Science **172**, 28 May, 1971.

18. The notion of the so-called "technological imperative," that is, the notion that new technological possibilities must be pursued wherever they may lead, is socially unacceptable. Thousands of projects which may be technically feasible are undesirable because of their harmful consequences (see AAAS Committee on Scientific Freedom and Responsibility, in *Science, Technology, and Society*, R. Chalk (Ed.), The American Association for the Advancement of Science, Washington, D. C., 1988. p. 249).

19. Engineering is said to have served society well by the standards society has set for itself.

20. J. Funder, Science **206**, 7 December 1979.

21. E. G. Mesthene, Technology and Culture **10** (No. 4), 489 (1969).

22. J. J. Baruch, Science **244**, 6 April 1984.

23. *The Nobel Prize Winners-Chemistry*, F. N. Magill (Ed.), Salem Press, Pasadena, CA, Vol. 1, 1901-1937, 1990; *The Nobel Prize Winners-Physics*, F. N. Magill (Ed.), Salem Press, Pasadena, CA, Vol. 2, 1938-1967, 1989; Vol. 3, 1968-1988, 1989.

24. T.H.Maugh II, Science **199**, 13 January 1978, p. 162.

25. J. I. Peterson and G. G. Vurek, Science **224**, 13 April 1984, p. 123.

26. W. Waidelich (Ed.), *Optoelectronics in Medicine*, Springer-Verlag, Berlin, 1982; J. A. S. Carruth and A. L. McKenzie, *Medical Lasers: Science and Clinical Practice*, Adam Hilger, Ltd., Bristol, 1986

27. M. W. Berns et al., Science **213**, 31 July 1981, p. 505.

28. B. R. Jasny and L. J. Miller, Science **260**, 14 May 1993. See also a special issue of Science (Science **260**, 14 May 1993) on biologically-based therapies.

29. D. C. O'Shea, W. R. Callen, and W. T. Rhodes, *Introduction to Lasers and Their Applications*, Addison-Wesley Publishing Company, Reading, Mass.,1978, p. 2.

30. Fiber optics came out of fundamental research in glass science, optics, and quantum mechanics and are still evolving. Optical fiber communication is driving the photonics revolution (A. M. Glass, Physics Today, October 1993, p. 34).

31. It has long been recognized that lasers can be used to encode information on materials that respond in an irreversible manner to exposure to high-density light. As a consequence of the coherence and relatively short wavelength of laser radiation, a large amount of information can be written into a very small volume of an appropriate storage medium.

32. Physics Today, June 1993, pp. 22-73.

33. M. Richardson, IEEE Journal of Quantum Electronics **QE-17**, 1598 (1981).

34. C. V. Shank, Science **219**, 4 March 1983, p. 1027.

35. J.- F. Nicoud, Science **263**, 4 February 1994, p. 636.

CHAPTER 8

The Cultural and Educational
Value of Science

Man's culture is rich in tradition and custom, myth and religion, wisdom and value, philosophical and artistic treasure, intellectual and technological skill. An inheritance -- rooted in the "tradition of intellectual honesty, which came to us from the Greeks; that of brotherhood, which we derived from Christianity; that of legal reason, which was the heritage of Rome; and that of tolerance which we were taught by Milton and Locke"[1] -- to be transmitted to and be transformed by future generations. To this heritage, Western culture has added *the scientific tradition and the scientific method.*

8.1. THE CULTURAL VALUE OF SCIENCE

Science is a relatively new element of man's cultural heritage. Although it is not as yet an integral part of the totality of human culture, its broad appreciation by society is necessary for society to be stable. There is a human reality in the cosmos, "a world with a logos," laws of individual and social existence along with those of nature, which the scientist often forgets. And there is more to science than its perception by the non-scientist as a threat to humanistic values. Segregation is as dangerous for the scientist as for the non-scientist. The "balkanization"[2] of science is as limiting to the scientist as is the fragmentation of man's cultural functions and their isolation from natural science to the non-scientist. The wisdom that humanity longs for, calls for a blend of science and the humanities, and a unified culture "in which the sciences describe a world which is alive with people and with feelings and the humanities describe a world in which the physical universe is not inert matter but rather is a part of the development of the human spirit."[3] Unfortunately, the spirit, the methods, and the basic concepts of (physical) science are still largely outside the interest and understanding of most members of society. Nonetheless, science has impacted man's culture in many and diverse ways. Below we give a few examples.

Science Provides Man with New Knowledge. Science, we mentioned in Chapters 2 and 5, continuously shapes man's views about the world and himself. It has uplifted human thought, comprehension, and knowing to new levels; it has opened up a world of boundless beauty and a horizon of limitless adventure; it has expanded the power and freedom of the mind, the potential of a warm heart, and the conscience of humanity. In so doing it has enlarged man's outlook on life. Man can now "see" as far as 300,000,000,000,000,000,000,000 km (3×10^{23} km), and as close as $1/(1000,000,000,000,000,000)$ km (10^{-18} km). Man can "look" as far as ten billion years (10^{10} years) and can follow events as fast as one millionth of a billionth of a second (10^{-15} s)! The methods of science equipped man with wings and unimaginable power.

Science, as well, impacted culture by continuously increasing man's semantic environment. Even a scientist who had lived at the beginning of this century would have difficulty communicating with his fellow scientists today without first learning their language. A physicist, for example, would have to learn what his colleagues mean by such terms as solitons, plasmons, gluons, quarks, black holes, quantum beats, superstrings, lasers, masers, excimers, buckyballs, and so on. Similarly, science is enriching the vocabulary of every person. A good part of society today knows and uses with ease -- albeit still with not sufficient understanding -- such words as laser, transistor, X-rays, electron, atom, nucleus, photon, quantum (quantum jump), fusion, fission, chain reaction, microwave, electrolysis, nylon, greenhouse effect, ozone layer, and so on. An increasing, but still small, part of society is familiarizing itself even with the content of science. For instance, the concepts of divisibility of matter, transmutation of energy and matter, and evolution are widely spread. Today, a judge may not know enough to understand the frontiers of nuclear physics, toxicology, biotechnology, or biology, but he may know enough to understand the issues of nuclear power, health and safety, and DNA fingerprinting.

Science Seeds Philosophy. Natural philosophy lies at the foundations of man's intellectual tradition. A philosophical bond unites the physical sciences and the humanities, a bond that becomes stronger as science becomes more abstract and, ironically, as it also becomes more relevant. The aim of philosophy and science is to reduce the variety of phenomena which we see to some principle, some simple fact, some law. In this, science has demonstrated for the philosopher the true complementarity of the deductive and inductive methods. Theory can neither be developed from facts alone, nor can it be deduced from first principles devoid of empirical content. It appears that the way of the philosopher is to begin with some presupposed and unverifiable postulates and to proceed deductively therefrom. Readhead[4] argues for a metaphysical realism: the external world exists independently of our knowledge of it and asks: "Are there foundations for knowledge, statements that we just know to be true, for which questions of further evidence simply do not arise, and which can then be used in their turn to justify other statements?" In his view neither Descartes (my own existence and the existence of God), nor Kant (the structure of conceptual thinking), nor Mach (direct sense experience) are credible. Verification of knowing insists Lafferty[5] must have a source which is not itself in knowing, otherwise it would be circular. Ultimately, the philosopher lacks the freedom and modesty of the scientist: the scientist has the experimental method and he only claims that his *knowing is evolutionary.*

The philosophical qualification of the concepts of truth, goodness, and beauty, as well as the philosophical answers to questions on the nature of life and the purpose of the universe necessitate the input of science, but are not science. Like the scientist, the philosopher wonders about the concepts of space, time, matter, order, force, law, conserved quantities, causality, verification, reality, complementarity, the modularity of the world.[6] He wonders, further, about the reduction of the macrocosmic universe to its microphysical constituents and the presumed simpler principles governing the behavior of the constituents (a macrocosmic universe which in one sense is made up of microconstituents, and which in another sense is not reducible to them). Indeed, the way of science -- synthesizing and harmonizing variety into a unified whole from which the behavior of the part is ultimately understood -- presents a model of thought for broader adaptation. As we have repeatedly pointed out, in the method of science independent measurements, observations, and facts, are correlated by formulating functional, empirical relations from them. These empirical relationships (the laws), in turn, are related by more general relationships, themselves unified by a theoretical frame of interpretation. Once this is achieved, the empirical laws follow as a deduction and thus the behavior of the particular is predictable. A synthetic, unifying approach: from the parts to the whole and from the whole to the parts (from the fall of an apple from an apple tree to gravitation and from gravitation to the laws of planetary motion and ballistics). And while science masterfully unravels the laws of regularities in the behavior of nature, it lets philosophy and theology make explicit the answers as to what nature really is, why it obeys the laws is does, or why it exists at all;[7] it lets them sharpen the contrast between *the metaphysical and the explanatory.*

If metaphysics "is the search for a statement subsuming the ultimate nature of all reality,"[8] science is not metaphysics. Science renounces such a claim; it deals with what is and not how it came to be or why. And yet, how can such a claim be valid in its totality? How can science be indifferent to the understanding of man's existence, and how can science be interested in unifying man's understanding of the behavior of the physical universe and disavow any interest in what this universe with such marvelous properties really is? "The sheer statement of what things are may contain elements explanatory of why things are," proposed Whitehead.[7] What science says about reality must be examined as to what it implies metaphysically. Indeed, then, science has affected the philosopher's metaphysics by forcing consideration of new elements on and new views about the universe and life. For instance, consideration of the mass-energy equivalence and the relativism of space and time as described by the theory of relativity, the structure and constitution of matter as envisioned by quantum mechanics, the nature of light and universal forces as claimed by physics, or the significance of the genetic code as illustrated by life sciences. How can philosophy not be compelled to wonder about the introduction into physics of absolute concepts and quantities such as the speed of light in vacuum, or the picture of an expanding universe sprinkled with black holes and neutron stars and yet as far as we presently know uninhabited, or the properties of transient particles presumed to be exchanged between the elemental entities of matter accounting for their existence and stability, or the complementary description of physical reality, or the statistical nature of our knowledge on the microcosmic and even in some instances[9] on the macrocosmic level? Further still, if regularities is what "makes life possible" and if through regularities man influences the world around him, how can philosophy not be compelled to wonder as to what the limits of our search for regularities are and as to how these limits limit what we can do or know?

By way of an example let us focus on just the concept of *time*. What is time? It is a mysteriously illusive concept of fundamental significance to all aspects of life. Morris[10] answers the question this way:

> "We all know what time is. It is what a clock measures. Nothing could be simpler than that. But when we begin to look at the question in detail, we discover that the topic of time is complicated indeed, that 'What is time?' is, in reality a whole series of questions. For example, what is this thing that we call 'the flow of time'? Does time always progress at the same rate? Is it like a flowing river, or is it the moment we call 'now' that moves from the present into the future? Is it conceivable that the flow of time could be stopped or reversed? Did time have a beginning? If so, how did it come into existence? Will time have an end? What was happening before the universe was created? Is time nothing more than a succession of events, or does it possess a kind of independent reality? Just what is this thing that a clock measures?"

Humanity has pondered over the concept of time for millennia. Ancient peoples (Babylonians, Chinese, Greeks, Romans, Aztecs, Mayas) believed that time was *cyclic* in character (historical events followed cyclic patterns which were somehow reflected in the nature of time itself).[11] The idea of *linear* time was introduced into Western thought by Judaism and Christianity where emphasis was placed on unique historical events presumed to have taken place at particular times. Indeed, Western culture embraced the idea of abstract, linear time. How can one speak of cosmic or biological evolution without the concept of linear time?[12]

Science does not answer the question "what is time?" It rather defines time operationally -- like Galileo, Newton, and Einstein did -- and provides evidence as to time's direction. The concept of absolute time (true and mathematical time, which according to Newton "of itself, and from its own nature, flows equably without relation to anything external"[13]) was followed by the concept of relative time (according to the special theory of relativity time measurements depend upon the state of motion of the observer, Chapter 3), and it by the concept of an intricate connection between time and space (according to the general theory of relativity the presence of matter creates gravitational fields that cause time dilation, Chapter 3). Science has identified scientific arrows (directions) of time. Physics, in particular, speaks of the arrows of time and in so doing it means that the world has a different appearance in one direction of time than it does in the other. The spontaneous flow of heat always takes place in the same direction, from the hotter to the cooler system. Similarly, the entropy of a system increases with time. Both define the direction of time. In the past the universe had less entropy and in the future it will have more; in the past the universe was more ordered than it is now (Chapter 3). The expansion of the universe, too, identifies an arrow of time in the macrocosmos. Matter in the universe was more compressed in the past and it will be more dispersed in the future. However, this method of defining the arrow of time vanishes in the microcosmos. At the subatomic level, individual particles can travel backward in time. There is no microcosmic arrow of time,[14] and the macrocosmic arrows of time still leave unanswered the questions: "does time have a beginning and an end?", "is there a boundary beyond which time does not exist?", "has the universe existed for an infinite length of time?", or "does time extend into the infinite past and into the infinite future?"[15]

Science is a Prerequisite of Wisdom. It appears that more often than not society is pulled by individuals and is equilibrated by the numbers. It is truism to say that both the individual and the society as a whole need more wisdom. The wise man, said Lao-Tse, looks into space and does not regard the small as little, nor the great as too big. Wisdom is a personal quality best acquired by quiet and persistent reflection and experience. It can be enhanced by the method and the content of science because science facilitates the basis for the development of critical thought, the correlation of knowledge, the choice among options, and the scrutiny of society's presumptions and value judgements. It helps make a wise and serene person *restless*! By increasing the power in the hands of man, science has increased the burden of the wise and deprived them of their detached isolation.

Science is a Model for Progress and Social Efficiency. Science can make democracy efficient by facilitating the means to do so (Chapters 2, 6 and 7), and by its own example. Society can profit by learning from science how freedom works in science and how freedom maximizes the efficiency of science. Democracy, after all, is secure as long as Newton's law of action-reaction applies to it. That is, as long as the force exerted by one part of society is balanced by the counter force exerted by the other part(s) of society -- as long as society is free to achieve that precious balance between the multidirectional forces exerted by its segments on one another. Science may, thus, help the appreciation of freedom in the efficient functioning of society by showing how freedom helps science coordinate and optimize its operations (Chapter 5).

   Science Cultivates New Attitudes. It enriches life by cultivating new and by reinforcing the value of old human habits.
   - *A habit of truth.* In the knowledge of man as in that of nature the habit of truth to experienced fact will uplift the level of civility and trust (Chapter 5).
   - *The value of self criticism.* Like every person who strives to excel, the scientist is under constant and ruthless self criticism. It is through this continuous reassessment and improvement that he achieves soundness in his experiments or in his theories. Although the world knows very little of this drama, it is not unlike the agony of the writer, the poet, the painter, or the mystic. They, too, go through many drafts, or stages; they, too, criticize themselves for they all strive for perfection and a glimpse of truth (Chapter 9).
   - *The necessity of discipline.* In the practice of science the development of the intellect is as important as is the development of the will. The practice of science imposes a discipline that shapes the character along with the stimulation of the mind. A discipline that breeds integrity (Chapter 5).
   - *The importance of openmindness.* The scientist must have an *open mind* and so must everyone else. Time and again the scientist has to reverse direction. In so doing, at times, he makes new discoveries, as, for example, when he expanded into relativity and quantum physics. One discovery follows another, each answering and each raising questions, each ending a search and each providing new instruments and ideas for a new one (Chapter 3). Proven scientific positions proved wrong no matter how great. Interpretations of observations, explanations of events and paradigms of theories have had alternative rationalizations and have always been limited,

never complete. This helps the scientist allow for error, strive for improvement, accept ambiguity, tolerate alternatives. Such attitudes are not a justification for relativism, but rather a lesson in humility, tolerance of alternative views, acceptance of the need to leave room for error. Society can profit a great deal from this kind of openmindness.

- *Receptivity to new ideas.* In science the habit of accepting various view points is encouraged. It maximizes one's options and teaches self-restraint. In science one fearlessly explores and defends his findings and ideas, and yet with restrain he learns to listen to the views of others even when they contradict his own. The content *and* the method of science accustom man to changing his beliefs.

- *The pride of collective achievement* (Chapter 5). The most socially significant events are characteristically the work of groups of individuals. There are few areas of human endeavor that exemplify this more or better than science. The reliance of one scientist on the work of his colleagues is a corner stone of science. It dramatizes the collective achievement and the collective responsibility in science. Meaningful participation in a broader effort and sharing in the common accomplishment instills pride in the individual whether in science or in other walks of life.

- The *recognition of originality.* Science has bred the love of originality as a mark of independence. The relation of originality to tradition in science has its counterpart in modern literature.

- *The significance of factual exactness.* The exactness of science cultivates the need for critically examined knowledge and it can give a context for our judgments.[16] It magnifies the need of mastering of the language as its indispensable tool.

- *The elegance of choosing.* Science has brought not just change but choice, often elegant choice. It presented man with more chances to choose. By multiplying man's choices, science forced on him the need to assume new roles of responsibility.

- *The challenge of exploring.* Science cultivates man's natural instinct for exploration, kindles his spirit of discovery, encourages the rush for new frontiers, and offers a method for such inquiry. The challenge of "pursuing unsuspected possibilities suggested by existing knowledge" is the basis of all progress in science, wrote Polanyi,[17] just as the discovery of electron diffraction was a confirmation of de Broglie's wave-theory of matter and the discovery of the positron was a confirmation of Dirac's theory of antimatter. The challenge of scientific exploration dares society become *a society of explorers.*

- *The anticipation of change.* This we have discussed in Chapter 5 on the distinct characteristics and principles of science. The anticipation of change induces a feeling of temporariness, affects man's planning, tilts his preference toward the short-range rather than toward the long-range aspects of his problems.

Science Provides New Means for Humanistic Studies. Science has fundamentally altered the methodologies employed by the humanities including those in teaching and research in these fields. It has influenced music (trend toward abstraction, computer music), the fine arts (abstraction, geometry and shape combinations), and literature (the concern with more analytical). Science has provided novel means to better understand people, and has challenged the humanities to balance the advancement of science and the growth of technology with human values.

The Rise and Fall of Civilizations. We have repeatedly said in this book that the recent advancement of science has been phenomenal and that the scientific frontier has been and remains endless. Yet throughout history civilizations had risen, reigned, and fallen. They always developed in local settings (Chapter 1). In the birth and transmission[18] of civilizations perhaps the most important factors are morale, confidence, deep belief, and enthusiasm. This way the ideas of a civilization remain fresh and young. The advancement of a civilization also requires that intellectual thought is efficiently translated into the language and understanding of the lay person. If the rise and the fall of all civilizations is an historical datum, is it not then proper to ask: *is the fall of the scientific civilization inevitable?* May be. But science remains constantly young and universally utilitarian. In contrast to previous civilizations which rose and fell in local settings and were identified with specific peoples and cultures, the scientific civilization is global and is thus not limited by natural or cultural settings. It is, however, limited in another way. It is evident that those who could be properly called scientists will never constitute more than a small fraction of the earth's population. Unless, then, the scientific culture becomes an integral part of man's total culture, the scientific civilization may wither away as all other civilizations did (see Chapter 10 on limits to science). Clearly, then, the cultural value of science must be transmitted more effectively and more coherently.

Many human institutions transmit the content, method, heritage and excitement of science, and articulate the role of science in the making of human culture. Among them, one institution -- *the university* -- has shaped science and has been shaped by science unlike any other. We shall, thus, focus a little on this awesome institution, but we shall first look briefly at the educational value of science.

8.2. THE EDUCATIONAL VALUE OF SCIENCE

Learning is wonderful and knowing is irreversible. Knowledge always overcame fear and the world is richer every day in things to know and in means and opportunities to do so. Recent knowledge, however, is prevalent, complex, and rapidly changing. The whole is incomprehensibly vast. Even the parts are beyond a person's grasp. Characteristically, E. Wigner (a Nobel laureate in physics) said in 1973: "I knew physics when I came to Princeton (in the late 1930's), I don't know physics any more, it's too large."[19] We become increasingly and uncomfortably aware that many of the useful things we learned at school are of little use to our careers, while other things which perhaps we had never learned at school are still as timely as ever, perceptively not affected by recent knowledge, qualifying in an almost mystical way the depth of our happiness. What, then, is the role of learning today?

We must, first, acknowledge the profound changes in the means and ways of learning that have taken place recently. There has been a proliferation of influences which shape our learning -- books, newspapers, periodicals, audio and video tapes, commercial and cable TV, new communication systems, the Internet. Technology has taken over many a function in both the formal and the general education process. Increasingly society and individual alike are involved in symbol, machine, and idea manipulations. They often fail under such circumstances to see the relevance of traditional teaching. A much larger fraction of the young is today in schools and an appreciable fraction of society at large is in some sort of training or education. These developments along with the revolution in the quantity, quality, and diversity of knowledge have changed the character of modern institutions of learning, their standards and goals, and the role of learning and learned persons in society. The dedicated teacher who once proclaimed that to teach he would build a trap such that to escape his students must learn, faces many and skillful competitors capable of building bigger and better traps for his students. He also faces an "educational" process which seems to have forgotten the most important

element of education: *the timeless and the truly important timely.*[20]
Ironically, then, modern man finds the golden treasures of knowledge
buried in a mountain of debris and himself powerless to control the
education of himself and his children. And yet, unavoidably, both the
formal and informal education influence and change us irreversibly. They
shape the quality of our character and thus the course of our civilization.
Today, more than in years past, we need to distinguish between *education
(παιδεία)* and *training (άσκησις)*, and to stress the need for both.

8.2.1. Education

Education is the social process through which man's cumulative
traditions and heritage are transformed and transmitted to future
generations, and the future is prepared. It is the process which triggers,
catalyzes, and speeds up the functions of the mind without intrusion. It
helps a person learn how to learn, appreciate what he had learned, find
happiness himself and not demand it from others. You cannot teach a man
anything, you can only help him to find it within himself, advised Plato and
Galileo. Education motivates a person to be curious, enables him to
distinguish the important from the trivial, focus on the fundamental and not
the merely useful, and embrace along with the topical and the socially
relevant the global and the truly human.

Education cultivates the non-specialized part of a person, breeds life-
long habits of a person's mind and character, aspires a person to
acknowledge the virtue of another, to possess his virtue itself. It aims at the
whole person whose attributes are freedom, moderation, sound judgement,
civility, *wisdom*, and *arete* (*αρετή*). Wisdom that stresses the need for
personal responsibility and arete that graces a person.

Such education is necessary for everyone including the scientist whose
narrow training may make him a specialist, but may deprive him of proper
appreciation of humanistic values and a broader perspective. It requires a
serious exposure to both classical literature and science. For without
science how can we truly appreciate, defend, and preserve what we value?
In a confusing and rapidly changing world, how can we value the value of
what we destroy? How can we prepare the young for living in a world that
uses deliberately science and technology as vehicles of social change if we
fail to teach them the scientific basis of every-day life? The place of science
in the world of values and facts must be understood by every educated
person. This is undoubtedly difficult, but it is perilous and immensely

unwise to do otherwise. Whether the unbalanced individual (and by extension the unbalanced society) has been the subject of comedy or tragedy is irrelevant: laughable or pitiful he is destined to demise.

Science, therefore, must find its proper place in the formal and informal educational systems, and classical education must be an integral part of a scientist's training. Training in a scientific field is neither synonymous with education, nor is it sufficient for the growth and the balance of a person. The scientist needs liberation from his narrowness by exposure to a broad liberal education, and the non-scientist needs proper instruction to overcome his fear of the *unapproachable* science by introduction of a science curriculum early in the educational system. Otherwise, the gulf between them will only widen!

8.2.2. Training

The preparation of a professional, a skilled expert for a career, is *training* (or extrinsic education). Through education we may better the world and prevent it from catastrophe, but it is through training that any of this can be done. Education can qualify, but it cannot replace training. It has been correctly said that the wheels of modern society cannot be turned without the specialists in the fields of higher learning. Today specialization is difficult. It demands curricula tailored to new knowledge and techniques designed to give the student the skills and know-how considered necessary by the peer professional. Training is demanded by those entering the institutions of higher learning, and by society which has jobs that must be done and be done well. Careers demand progressively more preparation and longer training. They are expensive to the trainees and to society. Consequently, training asserts itself and effectively squeezes education out of the formal curriculum. Certain types of specialization not only limit the education of the student, but deprive him of the ability to do "something else" later on. It is precisely for this reason that calls to shift from specialization to diversification have become louder in recent years.

There is still, as well, the area of science education for society at large, the ideal of a broad science-informed citizenry. Not one of amateur scientists, but one that can sense a feeling of participation in the excitement of science, pride in its accomplishments, and appreciation of science's contribution to man's culture. Science should be taught "with a full regard for the intellectual history and human values which went into the development of the sciences."[21] In this regard much has been accomplished by the Space Program and perhaps much has been lost by the Manhattan Project.

8.3. THE UNIVERSITY

The university is one of the oldest and most stable human institutions. It persisted through the centuries to this day under diverse and often adversarial conditions, reflecting (to some extent even until today) the views of its own faculty and students. By concentrating on study, the university secured its autonomy. Society has respected it for its product: *the educated student and the new knowledge.* While there has been a long struggle for the university to be a place of free discussion, by-and-large the university's success rests on its ability to draw a balance between academic freedom and academic responsibility. In so doing it has been able to keep alive its faith in intelligence, knowledge, and rationality and become a "citadel of reason," a spring of tradition devoted to learning and discovery, and a sensitive detector of social change. The university tolerated controversy, limited extremism, permitted expression of diverse and unpopular views, resisted the *status quo,* and served as a stable force in times of political and social unrest. A remarkable vehicle of change, it has itself changed the least -- until recently -- by the change it has caused. This conservatism has served well both the university and society. The university survived for twenty four centuries, since it was first established by Plato in Athens, Greece. Most of the 16th century European universities (for instance, those of Pisa, Padua, Genoa, Coimbra, Oxford, Heidelberg, and Prague) are still with us today. And so are the colonial universities of America founded in the late 17th to mid-18th century by the established churches that emerged into some of the most prestigious academic institutions of modern times [Harvard (1636), William and Mary (1693), Yale (1701), Princeton (1746), Brown (1764), Queen's College (1766, later Rutgers), King's College (1754, later Columbia), College of Philadelphia (1755, later the University of Philadelphia)].[22] Riled and pressured by the rulers of the day, the European university has nonetheless been a privileged institution. "Even under the Russian Czars the police were forbidden to enter the university, a tradition that curiously persisted through the Russian repression at Prague in the summer of 1968."[23]

8.3.1. The University is the Birthplace of Modern Science

The universities of Europe and North America have become global intellectual centers. The knowledge they created fostered much of modern science and technology. Two particular characteristics of the university have played a pivotal role in its unprecedented contribution to the development of modern science: *academic freedom and disciplinarity.*

Academic freedom is a remarkable human achievement for it embodies the notion that knowledge is preferable to ignorance, listening is preferable to silencing, and the new and the disagreeable and even the heretic have the right of expression and protection. "Academic freedom is the freedom of the scholar to search for truth, to reach his conclusions with intellectual honesty, and to retain his rights and privileges as scholar and teacher however unpopular his professional conclusions may be."[24] Because academic freedom has come to be identified with faculty problems such as tenure, some[25] prefer instead the term *intellectual freedom:* the right to have, explore, teach, and tolerate new ideas, and the right to discuss these free of social and political constraints. While academic (or intellectual) freedom is not unconditional, its "neutrality" along with science's transcendence of ideology provided the appropriate environment for science to grow.

The university has been and still is discipline[26] oriented. Its viewpoint has been and to a large degree still is the sum of the viewpoints of the separate disciplines comprising it. Departments along scientific disciplines have been the power bases within the university. The disciplinarity of the university embodies its traditional tendency to deal with pure, intrinsically motivated problems, generated and solved from within the disciplines. It allows the university to pursue pure and esoteric fields (e.g., high-energy physics), and explains the rationale for the university's standards of excellence: *excellence is what deepens the understanding in a particular discipline.* The standard is set from within the discipline.

The growth of knowledge and specialization along disciplines served society well. It is through the university disciplinary structure that the content, method, and excitement of science are transmitted, and it is through disciplinarity that the necessary knowledge and experts to work in the depths of well-defined disciplines can be generated. It is mainly through the scientific disciplines that an effective teaching of the history of science is accomplished and the scientific norms are taught and practiced. The disciplines best accomplish this in the proper execution of basic research.

Disciplinary research -- mostly in a field of "Little Science" (Chapters 3, 4, and 7) -- does more than advancing the field and producing new ideas and breakthroughs, it teaches the skills for, and the ethic and method in the practice of science. It generates new scientific leaders and new qualified workers for the industrial and technological society who carry over to the job not simply laboratory techniques and scientific know-how, but the method and the norms of science. Excellence in disciplinary basic research livens teaching and is a resource for any nation and for humanity. Competence in a discipline is frequently complemented by that in another. Collective disciplinarity solves problems outside the immediate field of interest of a particular discipline, as many scientific and technological developments since WWII have demonstrated.

8.3.2. The Inevitable Change

Never in the history of mankind has an institution influenced science as much as the university has had. In turn, the scientific revolution the university has helped create, has been changing the old institution itself. Some of the most obvious changes that have been brought about directly or indirectly by science are size, new missions, and multidisciplinarity.

Size. Today the university continues to maintain its pristine position as society's most distinguished institution of higher learning. It continues to shape society, to be society's foremost custodian of knowledge (a place where knowledge is generated, preserved, synthesized, assessed, and used), and a center of culture. Today's university is the *Alma Mater* of a large and influential fraction of society, and today's society is increasingly a product of the university and its outlook. This has broadened the university's influence and support base. The university's past patronage of princes and bishops has largely been replaced by government and industry. Most of the research universities in the United States are public institutions supported by state governments. While not all universities are universal or the same, almost all have a common characteristic: bigness. Some grew to the size of a town.

The modern university has become a complex and diverse institution (referred to by some as *multiuniversity*[27]), and a cultural regional center. It preserves knowledge (museums, libraries), offers aesthetic experience (arts, culture), and provides to communities university-based services[28] (maintenance of hospitals, clinics, testing labs, publishing companies, hotels, restaurants, bookstores), looks after young people, and entertains the masses. It has become, it is viewed, and it is run as a big business. Many universities are state-wide systems and others eye new institutional arrangements such as consortiums and alliances with industry and national laboratories as a way of expansion into bigger problems and into "Big Science" projects. Size has complicated the university's governance, and bureaucratization has pushed the university beyond its tradition. Like science, the university has gained and has lost a great deal by its bigness.

Beyond its Traditional Roles. The social processes that have shaped and are shaping science are doing the same for the university. Society pays for the university and demands a return. An increasing part of this demand is service. The university today has become essential for the well being of the individual and society in real, practical terms. Without a university degree one can make little advancement in today's society. Part of the service demanded from the university by society is therefore the education of the many. Today's university is indeed for the many, not, as in years past, for the elite. The egalitarianism, power, and bigness of the modern university bear little resemblance to its humble predecessor. Critics describe the modern university as an impersonal institution which sells instruction to the student, tailors its curricula and training to the needs and wishes of industry or the state, and functions as a knowledge industry and degree factory, a place where "students get credentialed."[29] Critics view today's university as an institution which has compromised and lowered its standards to accommodate social pressures and the numbers.

The university is also described as a place where modern society turns to for scientific discovery it needs to solve its problems. The university proved itself in helping society with its problems during and after WWII, why not ask it to serve more and better? And so the university has become a partner with government and industry in what has been called the "university-government-industrial complex,"[30] a good portion of which has to do with the military. The university has become a big place for "Little Science" and a participant with other institutions in "Big Science". Industry and government have gained access to the university, its faculty and students, its facilities, and even its governance. Their influence as big sponsors of research, educational programs, building facilities, employers of university-trained people, and users of university-generated knowledge has impacted not just the functions and directions of the university, but the university value system as well. The university had to adapt to the changing market place for its products (students and knowledge), the needs of industry, the planning of government, and the educational system demanded by society. This has affected and continues to affect the education and training of the student, the kind of inquiry pursued by the university, and the freedom and independence of the university. State, federal, and industrial support, meant that the university is constantly under the watch of outsiders. When research is supported by government, science in the university becomes a part of the political system, and when research in the university is funded by industry, the university researcher becomes constrained by the norms of the industrial sponsor. Harsh realities -- bigness, egalitarianism, regulation, bureaucracy, political governance, societal service, external controls, contingent support, changing attitudes and philosophies -- stress the university as it strives to preserve its past and adjust its future.

Multidisciplinarity. A consequence of the new developments, is a distinct esoteric change in the university structure: acceptance of multidisciplinarity. A recognition that while the growth of knowledge and specialization along disciplinary lines served the university and society well, there are dangers in such a narrow disciplinarity. It is feared that disciplinarity and concern for purity leads to remoteness and increased isolation of the parts from the main body of science, and that the narrow and arbitrary definitions of the fields of endeavor of the university reduce its potential to deal with societal problems which are multidisciplinary in nature. It has been argued, that the disciplinarity of the university not just hinders its ability to function as a "community rather than as an assembly of scholars,"[31] but it no longer reflects contemporary intellectual life. Interdisciplinary studies are needed, it is argued, because real world problems are not separable into disciplines and hence an interdisciplinary approach must be taught out of obligation to the young.[32] Society has problems -- thorny and messy problems -- and the young must know them and be equipped and motivated to solve them. The university must produce citizens not merely persons qualified in the fields of science.[33]

There are those who want the university to move into the city, be active in government, and help solve the progressive degradation of the quality of human life. Many university programs have been initiated in these and similar areas. Clearly, purity and fragmentation can be balanced with application and multidisciplinarity, specialization with diversification, acquisition of new knowledge with solidification and codification of existing knowledge, scholarship with research. These new roles, whether they are viewed as opportunities or as dangerous dilutions of the university, will further change the university. They will increase the influence of administration, redistribute power within the university, and readjust academic standards.

In the midst of change and its gusty winds generated by the advancement of science, the university must preserve the pivotal elements of its millennia-long tradition foremost intellectual freedom, effective education and training, sound research, and service to society. In the interest of science and society the university must remain intellectually open. To remain intellectually open, the university must limit its freedom by responsibility.

8.4. REFERENCES AND NOTES

1. M. Polanyi, in *Physical Science and Human Values*, Princeton University Press, Princeton, New Jersey, 1947, p. 131.
2. I. I. Rabi, *Science: The Center of Culture*, The World Publishing Co., New York, 1970, p. 92.
3. Ref. 2, p. 42.
4. M. Redhead, *From Physics to Metaphysics*, Cambridge University Press, Cambridge, 1995, p. 10 .
5. T. T. Lafferty, *Nature and Values*, University of South Carolina Press, Columbia, South Carolina, 1976, p. 40.
6. Particles that make up all of the physical world are supplied in a few "models" (the elementary constituents of matter), but in incredible numbers. All particles of each type are identically the same and indistinguishable from each other; a hydrogen atom is the same as any other hydrogen atom.
7. A. N. Whitehead, *Science and the Modern World*, The MacMillan Company, New York, 1926, p. 134-135.
8. C. Hinshelwood, Nature **184**, 12 December 1959, p. 1835.
9. In certain, albeit limited, instances the indeterminacy of the microcosmos enters the world of large scale phenomena. For instance, a single neutron can trigger a chain reaction in a quantity of fissionable material and cause an explosion just as a *single* X-ray photon can hit the nucleus of a living cell and produce a mutation. Both, the neutron and the X-ray photon, are subject to the acausality of the microcosmos. One may also note that the statistical laws affirm regularities based on the chance motions of a large number of entities (e.g., molecules in a gas).
10. R. Morris, *Time's Arrows*, Simon and Schuster, New York, 1985, p. 7.
11. Ref. 10, Chapter 2.
12. Ref. 10, Chapter 1.
13. D. Lindley, *The End of Physics*, Basic Books, New York, 1993, p. 59.
14. R. Feynman, *The Character of the Physical Law*, M.I.T. Press, 1967.

15. For further reading see Morris[10] and S. W. Hawking, *A Brief History of Time,* Bantam Books, Toronto, 1988.
16. J. Bronowski, *Science and Human Values,* Harper & Row, Publishers, New York, 1965, p. xiii.
17. M. Polanyi, Minerva V (Summer, 1967), p. 533.
18. See also a discussion on the transmission of a scientific civilization by M. J. Moravcsik (Bulletin of the Atomic Scientists, March 1973, p. 25).
19. E. Wigner, Science, **181**, 10 August 1973, p. 527.
20. B. Glass, *The Timely and the Timeless,* Basic Books, Inc. Publishers, New York, 1970.
21. Ref. 2, pp. 41 and 42.
22. R. Hofstadter and W. P. Metzger, *The Development of Academic Freedom in The United States,* Columbia University Press, New York, 1955, p. 115.
23. J. Barzun, Saturday Review, November 15, 1969, p. 23.
24. G. E. Pake, Science **172**, 28 May, 1971, p. 908.
25. See for instance, D. R. Goddard and L. C. Koons, Science **173**, 13 August, 1971, p. 607.
26. A discipline generally refers to a branch of knowledge and encompasses both teaching and research.
27. J. A. Gallant and J. W. Prothero, Science **175**, 28 January 1972, p. 381; Science **177**, 15 September 1972, p. 943.
28. R. G. Mawby, Science **238**, 11 December, 1987, p. 1491.
29. E. L. Boyer, The Chronicle of Higher Education, March 9, 1994.
30. F. Barnaby, in *Trends in Physics,* M. M. Woolfson (Ed.), Adam Hilger Ltd., Bristol, 1978, p. 51; K. A. Smith, Physics Today, February 1984, p. 24; C. Norman, Science **217**, 6 August 1982, p. 511.
31. Ref. 2, p. 93.
32. F. A. Long, Science, **171**, 12 March, 1971.
33. G. T. Seaborg, Impact of Science on Society XXII, January/June, 1972, p. 111.

CHAPTER 9

Where Science
Meets Religion

Human behavior has two irreducible aspects: the physical and the moral. Both complement man's search for comprehension and truth. Complementary descriptions though excluding one another, are nonetheless needed "to convey the rich possibilities flowing from man's relationship with the central order."[1] Science and religion in parallel and autonomous -- yet complementary -- ways stretch the limited perceptions of the senses to grasp the vastness of the physical and human reality, the beauty of the ideal, levels of experience and knowledge beyond the self. It is in the accommodation of this experience and this knowledge that man's actions are balanced and his judgements attain their value.

9.1. THE COMPLEMENTARITY OF SCIENCE AND RELIGION

Science tells us much, but never everything. The method of science is not the sole cognitive way to reality. Science is restricted to the world of matter, energy, space, and time, and deals with what physically exists. Science cannot -- and does not -- weigh or establish values. It renders no ethical judgements. It harbors no absolute or final truths, recognizes no eternal questions, claims no eternal answers, and has no substitute for self-correction and self-improving understanding. It openly modifies its position when the facts demand it. Science has no evidence for or against the existence of the soul, or the "reality of the miracle." There is no principle of love in science. This transcends science. Nevertheless, as Niels Bohr observed, although we can find nothing in physics or chemistry on consciousness, we all know that there is such a thing as consciousness, simply because we have it ourselves.[2] Indeed, in this regard, human consciousness constitutes a fundamental "experimental" datum. "The heart has its reasons of which reason knows nothing," affirmed Blaise Pascal.[3] There is a science of the soul of which science does not speak. And as Hannes Alfvén described, "it is scarcely possible to ponder the long chain of complications (of physical reality) without feeling a certain religiously colored reverence for the miracle wrought by nature; a miracle that is all the more fascinating for not being a simple juggler's trick."[4]

Although science is ontologically non-committal, is not ethically neutral. Science does not tell us whether it is right or wrong to kill, but provides for us innumerable and effective means to do so. Through its epistemology (Chapters 3 and 5) science provides a rational basis for ethical decisions by clearly stating the facts, the potentialities, and the options for good or evil. It induces the adjustment of religion to the condition of modern man and the evolution of values. It strips a person of superstitions and fears allowing this way a clearer perception of the universal order. Ethics as a moral system that defines duty and judges conduct are concerned with values and involve a critical examination of standards of good and evil, right and wrong, virtue and vice. The ethics of man cannot be based on science, but they cannot be divorced from science. Without science the ethics of modern man fall apart. "Conscience needs knowledge."[5]

Traditional ethics -- for instance, the Hellenic-Judaic-Christian ethic of the West -- are anthropocentric. Their significance rests mainly on the direct dealings of men sharing a common present, and on man himself. Traditional ethics normally involve the present and the regional for they are mostly concerned with man's contemporaries and neighbors. They are confined in their horizon with regard to the future and with regard to the distant. Under these conditions, writes Jonas,[6] besides moral will, the knowledge required to assure the morality of action was "not the knowledge of the scientist or the expert, but knowledge of a kind available to all men of good will."[6] Under these conditions human good is the same for all time, its realization or violation taking place at each time, the present being always its complete locus.[7] This is so no more. Science and technology had decidedly changed man's ethics by adding a new role of knowledge in morality, namely, by stretching modern man's ethics both in time and in distance. Man's previous ethics did not normally consider the global condition of human life and the far-off future of mankind. Man is now faced with what can be termed "delocalized" ethics. Ethics stretched in time and ethics spread in distance. The ethics of the "future neighbor," and the ethics of the "distant neighbor." Coexistive with traditional ethics, delocalized ethics impose on man new responsibilities -- the need to foresee and the need to guard against the threat to the image of man anywhere and at any future time.

Religion, on the other hand, deals with the person: man's life and death, his actions and behavior, his values and ethics, his faith and belief, his place in the cosmos, the ultimate meaning of existence, the purpose of the universe. It is a repository of "truths" one feels to be given and "facts" one accepts to be self-evident because they are part of him and because they exist in so many real, yet scientifically-speaking unprovable ways. Man -- even the man of today -- has to turn to religion to satisfy his need for faith. Religious faiths last and endure in contrast to ideological faiths (e.g., Marxism, Maoism), which fade away and die. Belief must be recognized "as the source of all knowledge,"[8] and faith as the fundamental source of personal strength. However, to the old question "who are we?" (τίνες δε ημείς;) not only man's ideological and philosophical systems but, also man's religious systems respond differently. There is plularism in belief in spite of the underlined uniformity in the teachings of the great religions of man. Let us then by way of example focus our subsequent discussion mostly on one of man's religions, namely, Christianity.

Christian theism prescribes the creation of the cosmos *ex nihilo* by an infinite, transcendent, omniscient, sovereign, and immanent God to operate orderly with a uniformity of cause and effect. It describes a universe with epistemological objectivism, and with an objective reality to be discovered by man. Man has the capacity and indeed the obligation to discover the cosmos, for the God who created the cosmos also created man in His own image *(imago Dei)*. Man, thus, possesses the intelligence and the ability to unravel the laws of the cosmos. Science, wrote Polanyi, is guided by the urge to make contact with a reality which is felt to be there already.[9]

The Christian God of the universe is also a *personal* God, the father of each and every individual, who constantly cares for and communicates with everyone. Through Him each person is united with all other human beings (his brothers) and the entire Universe. This is indeed a united and revered Universe. A cosmos whose universal stability and intimate bonding is mediated by the force of love. One where man's ethic and spirituality are transcendent because they are based on the goodness of God. A universe where man and his behavior can only be understood in relation to God, and where *the presence of the image of God in all entails the sanctity and unconditional dignity of all.* To quote Lossky, "each person is an absolutely original and unique aspect of the nature common to all."[10]

Christianity teaches the universal awareness of God, acknowledges the principle of non-contradiction, has a linear view of history, emphasizes the value of truth, and stresses that all truth comes from God and is thus whole. This unity of truth underlines the mutual consistency of all true propositions, and presupposes true complementarity among the various types of knowledge (scientific, philosophical, intuitive, empirical, revealed). Without complementary knowledge the truth would not be whole. And without such outlook science would not have emerged the way it had. "Philosophic speculations and monotheistic theologies were the necessary precursors of the weaving of the fabric of modern science,"[11] maintained Conant. Science and technology would not have come about without the Judeo-Christian tradition, believed Whitehead.[12] The tradition that Whitehead speaks of is rooted in the belief that an intelligent Creator created the universe, laid down the laws of its operation, enabled man to understand the universe, and gave him the reason and the urge to observe it and to experiment with it. What a miracle! The man is empowered on the one hand to explore the beauty and the order of the cosmos, and on the other hand to become one with God -- and the choice is left up to him!

In general, religious concepts are older than those of science. The basic teachings of the great religions, in contrast with those of science, have changed little over the centuries. Unlike the scientific law which can be broken, changed, amended, or abandoned, the moral law of religion is unbreakable and unchangeable. It does not lose its validity by being broken.[13] The wisdom and the belief of religion bears no similarity to the "standard" of science. Even the language of religion differs from that of science. For instance, while an analogy can be drawn between the *parable of religion* and the *model of science*, there is no place in science for the parables. And yet, how can one express the truth of the parables in another or a better way? Each parable describes a paradigm that complements the rest by offering a particularly unique perspective on truth, on man and on man's behavior. As in science, acceptance of one paradigm does not mean giving up another. In science for instance the scientist accepted the paradigm of the *quantum* without giving up the paradigm of the *wave*; he needs both for a more complete description of physical reality. Similarly, in religion the Christian ethic is more complete with the paradigms, for example, of the parables of the Lost Son, the Good Samaritan, and the Talents, than by just the paradigm of any one of them.

Both science and religion are hindered by expression and concept, yet both are able to convey their most fundamental aspects with magnificent simplicity. How simple and elegant and yet how fundamentally profound is the knowledge embodied in the laws of Newton's equation of motion ($F = ma$), Einstein's mass-energy equivalence ($E = mc^2$), and Maxwell's equations describing the nature of electromagnetic waves. Similarly, in religion, how elegant and real are the messages of the parables and Christ's Sermon on the Mountain.[14] There is a majestic beauty in the simplicity of mathematical descriptions of physical reality and in the design and execution of simple but critical experiments.[15] Similarly, there is unique elegance in the simplicity of the message of the man of religion.[16]

Science and religion speak of faith and doubt. In both the critical steps to knowledge come hesitantly from within. "My intuition tells me so," responded Enrico Fermi when proof was demanded of him in a scientific discussion.[17] Peoples (Gentiles) who have not the law do by nature what the law requires, they are a law to themselves,[18] taught Saint Paul. It is by logic that we prove, but by intuition that we discover, affirmed Poincaré. Whether in science or in religion, discovery is a flash of insight, a gift to the prepared mind or to the prepared heart, and it proceeds from faith. "The discovery of a fundamental, verified law of nature is an inspiration of God," said others.[19]

The essence of our understanding of nature in science is as personal and many-sided as is the grasp of the divine in religion: always difficult, incomplete, agonizing. For whether we struggle for knowledge of the physical world in science, or for comprehension of the universal truth in religion, we are handicapped by our own limits. We are constantly experiencing the discomforting uncertainty of shadows. Ironically, the essence of man's knowing rests within the realization of the extent of his ignorance and limitations. Nonetheless, in both science and religion, the level and quality of one's knowledge enhances his ability to rid himself of fear and be free.

A true complementarity between science and religion requires mutual respect and accommodation. It is necessary that science and technology not be looked upon with contempt and fear as the forces of evil to be stopped before they destroy humanity or as the enemies of religious beliefs. After all, Nature and God exist independently of what man's science and of what man's religion say about them. Religion must accept the proven facts of science and the science description of the physical world, and science must stay away -- as science -- from human values and the areas of knowledge that transcend it. Religious beliefs should not be based on what science is, or on what science is or is not able to do. By its own nature scientific knowledge will change tomorrow. Is religion going to change its beliefs every time a new scientific discovery comes along? One's faith must stand the light of knowledge acquired by modern man and continue the same -- a guiding and immutable value. Christ used analogies from nature in His teachings, but he solicited neither the help of Greek science nor the support of the astrologers of Mesopotamia, observed Tsirintanes;[20] nor did He feel threatened by them. Past conflicts between science and religion can largely be traced to religious perceptions of physical reality that were not compatible with scientific findings. Such conflicts, whether in astronomy (the heliocentric system),[21] in geology (the age of the earth),[22] in physics (the uncertainty principle, determinacy, causality and free will),[23] or in biology (the question of origins),[24] led to serious misunderstandings and impacted negatively on both. And while the rationality of the universe as revealed by science is consistent with religion's contention of a created objective reality, and the awe and mystery of modern scientific discoveries and postulates (as, for example, the Big Bang theory and the postulation of a cosmic beginning) resemble more and more the awe and mystery traditionally found in religion, any attempt to drag religion into the field of science and science into the domain of religion is as perilous today as it had been in the past. Religion must be free from the bonds of the evolving science. Science and religion can best serve man by been complementary.

9.2. THE CONTACT OF SCIENCE AND RELIGION

Where, then, do science and religion meet? At the outermost edges and the beginning of the universe? Perhaps. In man? Unquestionably yes. In him they meet each other face to face. They, in fact, not only meet in him but they also *depend on* him. Both the physical and the moral law are meant for man. Only man can understand the workings of physical reality and only he can uphold values.

In man, science and religion are glorified and are condemned. In him they are both confronted with their past and are forced to ponder over and assess their future. In man they face the beauty they have helped him discover and simultaneously recognize perceptions of themselves that should die. In him they see the good they had helped him accomplish for his mind, spirit, and material needs, and are embarrassed by a litany of false religious claims and unfounded scientific promises made in their name that left man perplexed and resentful. How can we ignore the suffering inflicted on man and his societies by the help the institutions of science and religion granted to the *status quo* of dictators, oppressive regimes, and overzealous fanatics? Science and religion demand from their practitioners honesty, accountability, and a continuous responsible stewardship of the physical and the moral world. Clearly, then, the focal point in the contact of science and religion is the *Faithful Scientist, for **only where faith exists can science exist with faith.*** And just as the method of scientific inquiry differs from scientist to scientist, so is with faith in religion; it, too, differs from individual to individual.

As people in other professions, many scientists have been and are non-religious, or agnostics, or atheists. But there have been and there are today many other scientists who are deeply religious persons (see also Chapter 6). Many of the founders of science, such as Galileo, Leibnitz, Newton, Faraday, Maxwell, Cauchy, Thomson (J. J.), Planck, Einstein, Boyle, Pasteur were religious persons and some devout believers. Michael Faraday's "religious feeling and philosophy could not be kept apart; there was an habitual overflow of the one into the other," said his agnostic contemporary John Tyndall.[25] "Man infinitely transcends man, and without the aid of faith he would remain incomprehensible to himself,"[26] was Pascal's conclusion. "Science without religion is lame, and religion without science is blind,"[27] believed Einstein.

There are scientists who laugh at the belief of Christ's resurrection, and there are scientists who truly wish their loved ones to be as happy as those who first saw the resurrected Christ! There are scientists who deny the existence of God and who desperately search for substitutes (Chapter 6), and there are scientists who see God and His grace daily in their lives.[28] There are as well scientists who "profess ignorance in religious matters"

not because this is what science says, but because they have not been aware
of religious teachings or because they were trained to refrain from
expressing opinion and take positions on matters they are not certain about.
There are also scientists who wonder how the scientist who is devoted to
discovering the truth can reject another way to the truth simply because his
instruments or his mathematics cannot measure or formulate. There are still
those to whom God is real not for scientific reasons, but for a million
others: "I find a need for God in the universe and in my own life" (A.
Schawlow);[29] "I believe in, pray to, and worship God" (E. Nelson);[30] "The
exquisite order displayed by our scientific understanding of the physical
world calls for the divine" (V. Kistiakowsky);[31] "The concept of God...
helps us make decisions in the right direction. We should be very different,
I fear, if we did not have that concept" (E. Wigner).[32]

The devoted scientist and the saint of religion are dedicated to a way
of life rather distinct from that of the "common" person. They value truth
and they know the value of discipline in the practice of science and
religion. They transcend local, national, and institutional boundaries. They
are grateful for what they perceive and they are humble before the world
they marvel. Whether they believe in God because of the beauty and
reliability of the laws of nature, or because of the breakdowns in the laws
of nature (the miracles),[33] they constantly live at the edge of mystery. The
marvels of the physical and the spiritual worlds reinforce each other in a
mutual interplay of a perennial pursuit of knowledge. There is infinite
physical space for the scientist to move in as a scientist, and there is even
more spiritual space for him to move in as a person of faith. The limits are
different as are the prerequisites, but the ability and freedom to move in
both the physical and the spiritual space is a uniquely precious gift to him.
The desire to find the truth noted C. P. Snow is "itself a moral impulse,"[34]
its acquisition a perfection of man's existence. Whether they search for
knowledge in the physical world or look into another person for knowledge
of themselves, they learn to respect man and his accomplishments.
Generosity and integrity are basic elements in the method of science and in
the way of religion from which grows a sense of dignity that *links the ethics
of science and the ethics of religion.*

Scientific and religious faiths are similar and both presuppose doubt.
The scientist knows that science itself is based on faith (Chapter 6). Every
theory and every "fact" in science is an approximation however logical and
wonderful! The scientist knows, as does the faithful, that to see wonderful
things he must travel far and travel at times alone! He knows, too, that he

cannot define the limits of knowledge and thus he is content not to define God. Time and again he ponders over Christ's words that Eternal Life is the knowledge of God[35] and infers that *knowing must be the essence of eternity.* And though the window of science is narrow and the roots of faith atrophic, the magnificent wonders of the microcosmic and macrocosmic universe unraveled by the scientist -- far more magnificent than any prophet has ever perceived -- partake of the light of eternity. They refine man's idea of nature and of God and reassure his belief that life beyond the grave is real. Its most distinct characteristic -- becoming one with God -- been the ultimate miracle of all.

9.3. TOGETHER FOR A HUMANIZED EARTH

Inextricably, the big questions that confront man today bring science and religion together and both closer to man although on the surface they seem to be doing just the opposite. Recent scientific advances pose serious ethical questions which shake the traditional stability of religious institutions. Conversely, the narrow beliefs and rigid doctrines of some of today's religions often stand in the way of the pursuit of knowledge. *This stress is nonetheless a symptom of an agonizing and continuous struggle for mutual accommodation and reinforcement, an integration of man and his faculties, and the result of the enormously complex consequences of the social and ethical roots of the advancement of science.*

The enormous power that science has placed in the hands of man cannot be for good if man's stature is reduced to that of animals, or if religion is reduced to the level of opinion and man confines himself to his legal systems and neglects the transcendental ethic of religion. For the realms of modern biology (especially genetics and the human genome)[36] and biomedicine[37] have added new strength to the hurricane of change: control of man's biological evolution by changes in the genetic characteristics of the unborn and by human cloning;[38] modification of human (ethical?) and societal behavior; the definition of and right to life and death;[39] the power to determine the future genetic heritage of virtually any species. Each of these and other similar issues, but foremost the possibility of man taking control of his own evolution, carry man beyond the bounds of human tradition and all of his former ethics. They also carry the scientist beyond the traditional scientific norms and make it difficult for him to claim that the knowledge he produces is by definition good, especially when the science he does has shifted to a business-type activity largely in the hands of big drug companies.[40] The scientific/religious debate is quickly shifting from physics to biology. From "Does God play dice?" to "Should man play God?" In answering the latter question, man must decide which God to play.

New and extraordinarily ferocious problems. New additions to the long list of problems -- foremost world population and the possibility of a nuclear war -- already delivered to man by the hurricane of change. All these diverse and important issues have one thing in common: *they all transcend science and they all transcend religion*. In their confrontation science meets conscience. Science and ethics become inseparable, demanding an ethic for change itself. What a difficult and yet what a wonderful moment! Science and religion realize, almost unwillingly, that they have no choice but to *join hands for a "humanized" earth*.

9.4. REFERENCES AND NOTES

1. W. Heisenberg, *Physics and Beyond*, Harper and Row, Publishers, New York, 1971, p. 89.
2. Ref. 1, p. 114.
3. A. Krailsheimer, *Pascal*, Oxford University Press, Oxford, 1980, p. 78.
4. H. Alfvén, *Atom, Man, and the Universe*, W. H. Freeman and Company, San Francisco, 1964, p. 103.
5. J. E. Smith, American Scientist **68**, September/October 1980, p. 554.
6. H. Jonas, *The Imperative of Responsibility*, The University of Chicago Press, Chicago, Illinois, 1984, p. 5.
7. Ref. 6, p. 6.
8. M. Polanyi, *Personal Knowledge*, The University of Chicago Press, Chicago, IL, 1968, p. 266.
9. M. Polanyi and H. Prosch, *Meaning*, The University of Chicago Press, Chicago, IL, 1975, p. 63.
10. V. Lossky, *In the Image and Likeness of God*, St. Vladimir's Seminary Press, Crestwood, New York, 1985, p. 107.
11. J. B. Conant, American Scientist **55**, 311 (1967).
12. A. N. Whitehead, *Science and the Modern World*, MacMillan, New York, 1926.
13. K. Lonsdale, Nature **193**, 209 (1962).
14. Mathew **5**, 1-12.
15. Such as those with which Pasteur proved that life comes from pre-existing life, or those that led to the discovery of the neutron (Chapter 3).
16. Such as in the message of the man of faith who took part in the May 1, 1990, parade in Moscow and raised his sign for Gorbachev to read: "Christ is risen Michael Sergenev," the message simply read (NBC News, 6.30 p.m., May 1, 1990). Or like the story of the

two monks who lived together for a long time without ever having quarreled and who one day decided to do so to find out how people quarrel. In search for something they could have reason to quarrel about, they found that their only possession was a bucket of water. They decided to each claim it his own and fight over it. So, when one of the two claimed the bucket to be his, the other, after a period of silence, gave his simple answer. "Well, you say the bucket is yours. Have it!"

17. M. Wilson, The Atlantic, September 1970, p. 101.
18. Romans 2, 14.
19. A. Einstein, 1932, quoted by H. Margenau, in Cosmos, Bios, Theos, edited by H. Margenau and R. A. Varghese, Open Court, La Salle, Illinois, 1992, p. 62.
20. A. N. Tsirintanes, Knowing Where We Are Going, Cassell, London, 1977.
21. For instance, the Catholic Church lost prestige by its insistence on a geocentric solar system as "an article of faith" in contradiction to the facts of science which showed the sun in the center of the solar system and the earth moving around the sun, and by its persecution and condemnation of the scientists who presented the facts as they saw them, notably Galileo Galilei. Only recently, the Catholic Church corrected its mistake.
22. The age of the earth as estimated by science is billions of years, much longer than church's own estimates. Early books in geology estimating the age of the earth brought on the wrath of the church.
23. For example, the long discussion about the perceived relation between the uncertainty principle in the quantum mechanical description of the microcosmic universe and causality, determinacy, and free will. While quantum mechanics replaced determinism at the atomic level by statistical causality, large scale events (the events of every-day life and the macrocosmos) are strictly deterministic. Causal relationships whether deterministic or statistical, are governed by definite laws. Whether natural laws determine events completely or statistically has no bearing on the question of free will. Quantum physics has nothing to do with ethics.
24. For instance, the debates concerning the universality of the genetic code and man's special place in the cosmos. The former is not evidence against the latter.
25. R. J. Seeger, Physics Today, August 1968, p. 30.
26. B. Pascal, in Ref. 3, p. 52.
27. A. Einstein, quoted by C. Chagas, Science 207, 14 March 1980, p. 1159.
28. H. Margenau and A. Varghese (Eds.), Cosmos, Bios, Theos, Open Court, La Salle, Illinois, 1992.
29. A. Schawlow, in Ref. 28, p. 105.
30. E. Nelson, in Ref. 28, p. 75.
31. V. Kistiakowsky, in Ref. 28, p. 51.
32. E. Wigner, in Ref. 28, p. 131.
33. J. D. Barrow and J. Silk, The Left Hand of Creation, Oxford University Press, Oxford, 1983.
34. C. P. Snow, Public Affairs, Charles Scribner's Sons, New York, 1971, p. 192.
35. John 17, 3.
36. DNA is recognized as the genetic material in all organisms ranging from viruses to humans. It has been found [e.g., see A. G. Motulsky, in Science, Technology, and Society, R. Chalk (Ed.), The American Association for the Advancement of Science, Washington, D. C., 1988, p. 152] that restriction enzymes cut DNA at specific base sequences, different restriction enzymes each splitting DNA at different specific sites.

DNA that has been split by a given restriction enzyme can combine with any other DNA molecule cut by the same enzyme. It is, therefore, possible to join DNA molecules from different sources to reproduce so-called "recombinant DNA" consisting of parts of DNA from different species (species barriers can therefore be crossed), and it would be conceivable to produce man-made chimeras by the introduction of selected nonhuman material into developing human embryos (L. R. Kass, Science **174**, 19 November 1971, p. 779). The term genetic engineering refers to research advances over the last two decades or so which have been made by recombinant DNA (gene-splicing) techniques. Recombinant DNA methods can be used for a number of different purposes (gene understanding and functioning, diagnoses of diseases, production of drugs, medical treatment) each of which must be assessed ethically on its own merits. The term negative eugenics normally refers to improvement of the individual in particular cases, while the term positive eugenics refers to improvement of the entire human race. Both raise troubling issues, especially the latter (see, for instance, L. R. Kass, Science **174**, 19 November 1971, p. 779; Science **264**, 17 June 1994, p. 1685; P. Ramsey, *The Fabricated Man, The Ethics of Genetic Control*, New Haven, Yale University Press, 1970). Through genetic engineering man will be able to make changes that can be transmitted to future generations and this way he can create new possibilities and thus establish new norms for human health and fitness. Kass wonders if under these circumstances human parenthood could be kept human!

37. Through advances in biomedicine man now extends life, controls human birth and fertility, and has power over the generation of human life and modification of human material.

38. Announcement of cloning of an adult mammal (a sheep, born on July 5, 1996) was made in February 1997 (I. Wilmut, A. E. Schnieke, J. McWhir, A. J. Kind, and K. H. S. Campbell, Nature **385**, 810 (1997)).

39. It is worth focusing on this issue as an example of the multidimensional impact on society and its ethics of the advancement of science. Here is how B. G. Zack, [*Science, Technology, and Society*, R. Chalk (Ed.), American Association for the Advancement of Science, Washington, D. C., 1988, p. 98] describes the situation:

> "The Congress of the United States has asked medical science to tell it when human life begins...., to tell it when human life begins, so that it may know when to define its ending as a crime.... The law wants to know if the zygote, embryo, and fetus are human lives because it wants to know if these entities are entitled to the same rights and protections which the community has agreed to confer on human beings who have already been born. The issue is thus not whether the zygote, embryo, and fetus are human lives in a scientific, definitional sense.... The issue is at what stage of development shall the entity destined to acquire the attributes of a human being be vested with the rights and protections accorded that status. It is to the moral codes of the people that the law must turn for guidance in this matter, not to the arbitrary definitions of science.... To ask science to define human life in scientific terms for use by the law in moral terms is a travesty of both honorable traditions."

40. J. Cohen, Science **275**, 7 February 1997, p. 767; J. Enriquez, Science **281**, 14 August 1998, p. 925.

CHAPTER 10

Limits of and to Science

The unknown is infinite and the scientific frontier is "endless," but knowing is limited. Natural, practical, and societal barriers limit man's knowledge, understanding, and skill -- and consequently science. There is no end to science, but there are limits of and to science.[1] The limits of science we term *intrinsic or natural,* and the limits to science we term *extrinsic* or *socio-political.*

10.1. LIMITS OF SCIENCE

The natural limits of science are difficult to define and quantify for this presupposes knowledge of the nature and the extent of the limits which man does not possess. Nonetheless, areas connected with the intrinsic or natural limits of science can be identified, and some are discussed below.

The Nature of Knowing. Scientific knowledge is presumed to be objective knowledge. Yet, the objectivity of science has been questioned on the premise that neither the facts of science can be truly objective nor scientific detachment possible, but rather they are both conditional. All knowledge, scientific knowledge included, it is argued,[2] is *personal,* and consequently imperfect. Even in the exact sciences, Polanyi maintains, "knowing" is an art of which the knower is a logically necessary part. If "the concept of 'meaning' can be fathomed only in relation to the self," then, "meaning is intuitive," for the concept of the self cannot be defined explicitly, but intuitively.[3] Hence, the observer influences the observed, and -- conversely -- the observed and the perceived influence the observer and condition the thinker. Inescapably, the scientist is limited by mental conditioning and the perceived nature of nature. Scientific knowledge, many argue, is ultimately subjective knowledge.

The scientist is also limited by the tools and the concepts he uses. He is limited by experimental possibility, by the state-of-the-art of technology, and by the method available to him. However sophisticated the experimental method is, it principally relies on the measurement of two quantities: charge and light. Similarly, however powerful the concepts used by science are, they remain limited approximations based on the acceptance of defined entities. These entities are always insufficient and their *essence* unknown. Whether we know by induction, by analogy, or by deduction, we are limited by the specifics of the part, by the danger of the analogy, and by the implicitness of the generalization.

There is yet another component to this limit: that of language and semantic environment. Language, says Martin, now threatens to become a brake on man's future.[4] It confines the description of not just emotion and feeling, but also thought. Language and the proliferation of man's semantic environment limit man's comprehensive prowess.

The Nature of the "Laws" of Nature. As we discussed in Chapter 4, an implicit assumption in all physical knowledge is that the laws of nature apply always -- in the past, the present, and the future -- and everywhere. This universality and transcendence of the applicability of the physical law over times and dimensions that vary by factors of over 10^{40} (Chapters 1 and 4) is magnificent and powerful, but it is not without a limit. The universal applicability of the physical law is implicit (there is no evidence to the contrary), not proven. For example, we do not know with certainty that the laws of nature as they have been established on earth over the last three hundred years or so were applicable at the beginning of the universe. As we discussed in Chapter 4, the laws of the macrocosmos have been derived from light signals received from distant stars which do not now exist. Light travels, we assume, from the past to the future and thus the light signals tell us about the past, but not about the present state of the universe. It is common in science, however, to detect signals due to events or entities long gone, not only in the macrocosmos but also in the microcosmos. For instance, the time interval that separates the initial interaction of a charged particle with an atom or a molecule and the signal detected by the scientist which he attributes to that initial interaction and that particular particle is vast on the microcosmic scale although infinitesimally small on the macrocosmic scale.

The generalizations established by science -- such as the uniformity of nature (matter is composed of the same particles and elements everywhere, proteins in different species have a common character, cellular structure and cellular reproduction are of general occurrence), and universality of its laws (electromagnetic radiation travels with the same speed everywhere in empty space, the nature of the forces are the same everywhere, the genetic code is universal) -- carry with them a sense of completeness, and, thus, an associated limitation.[5] Such a view is actually not recent. For instance, Lord Kelvin toward the end of the 19th century -- and others since -- expressed this sense of completality. The knowledge that remains is "an infinite detail" they said or implied. Science has proven their prophecies wrong! The generalizations of science are *not* final, absolute, or complete descriptions of reality. They are beautiful summary statements of incomprehensible significance, yet they are temporal expressions of the behavior and appearance of only a small part of reality, bound to change with time. While the claim of universal validity has its completality aspects, there exists infinite knowledge and experience outside of present-day science. Not just new facts, but *new views* of reality. The limit does not rest with an achieved completeness in our knowledge of the physical universe, but rather with the size of knowledge and complexity of the next level of knowing and the new view and understanding of the universe that can be achieved. And while new methods and approaches will undoubtedly allow better understanding of complexity and its "global" behaviors,[6] still the better understanding of complexity will unquestionably confront us with more complexity. "Complete" is not synonymous with "all-encompassing." Classical physics is "complete" in that it could never be proven false within its own framework of concepts, but it is not "all-encompassing."[7] It does not, for example, encompass quantum and relativistic effects.

There is still another limit of the physical law: *life modifies the laws valid for inanimate nature.* A number of physicists[8] have commented on the relation of life to present-day physics and the extension of the latter into the science of life. Clearly, the laws of physics will have to be modified if they are to account for the phenomena of life. To quote Wigner, ".... the present laws of physics are at least incomplete without a translation into terms of mental phenomena .... The view given here considers inanimate matter as a limiting case in which the phenomena of life and consciousness play as little a role as the non-gravitational forces play in planetary motion, as fluctuations play in macroscopic physics,"[9] or, for that matter, as gravitational forces play in atomic and nuclear physics. And Wigner, in his traditional symmetrical approach, goes on to say that "since matter clearly influences the content of our consciousness, it is natural to assume that the opposite influence also exists, thus demanding a modification of the presently accepted laws of nature which disregard this influence."

Science is limited as well by the laws of nature themselves. For instance, science cannot violate the universal conservation laws. Neither can science experiment with cosmic phenomena; one can hardly argue that astronomy deals with control experiments. Nor can science deal with extremely energetic particles under laboratory condition; for the study of these science has to patiently observe nature.

Complexity. Large segments of science are incapable of dealing with inherently complex systems. Complex sciences, such as biology, are too difficult, if not inaccessible, to rigorous scientific scrutiny such as exercised in physics and chemistry. Quantum mechanics, for instance, finds it difficult to penetrate the essence of these areas although many attempts have been and are being made such as in computational and structural biology.[10] Science may thus be limited by the range of abstraction and complexity which human thought and method can cover.

Complexity also imposes a limit on science's ability to logically synthesize the elements of the macroscosmic and microcosmic universe it had unraveled and reconstruct reality. The frontiers of the advancing science have been mainly associated with an increase in resolving power -- inwards toward the elemental constitution of matter and outwards toward the galaxies -- but, having resolved the parts, we still cannot understand their re-association to reconstruct reality.[11]

The Vastness of Knowledge and the Phenomenal Expansion of Science.
It has been correctly said that man can be *any* of *many* things, but he can
never be *all* things. The recent explosion of scientific knowledge resulted
in the well-recognized fragmentation of science. Each scientist, by
necessity, narrows his field of concern. The narrowness of this
specialization, impairs the efficiency of science and carries with it the
danger of limiting the proper fertilization of the branches of science.
Isolation of the parts of science engenders the risk of them becoming
inconsistent with each other, unaware of relevant advances in other fields,
in spite of the extraordinary recent communication systems. Fragmentation
and growth limit the intellectual discipline of science. Furthermore, elitism
and drive for excellence in increasingly esoteric science widens the gap
between science and the rest of society. The growth of science is self-
limiting.[12]

And science has its fashions. Scientific fields become "complete" (for
instance, classical thermodynamics and Newtonian mechanics), or are
vigorously reborn with the advent of new advances (for instance, optics
with the discovery of the laser), or reach a state of complexity and have to
await a new breakthrough for their furtherance (for instance, the
understanding of nuclear forces), or while fashionable in one area of
science in time become the concern of scientists in another (for instance,
nuclear, atomic, and mass spectrometries once areas of physics
subsequently became active fields in chemistry, engineering, and biology),
or simply appear at a given time and captivate a large portion of scientific
activity by virtue of their novelty (for instance, clusters, C_{60}, high-
temperature superconductors). The fashions in science complicate the
balance of scientific knowledge and influence the proper teaching of
science. Although it is only natural for fashionable areas to excite
researchers and students alike, the fashionable often hinder the traditional
functions of science and intrude in the proper education of the graduate
student who at times knows amazingly well complicated and fashionable
subjects but lacks the elementary understanding of basic science.

Man's Limited Faculties. Perhaps the most distinct limit of man's
intellectual capacity is his inability to define his own limits. He is
obstructed by the abstraction and the complexity the definition itself
carries. Man's unlimited thought is limited. Increasingly, new paths are
reminiscent of old; straight and clear they seem to be initially but they
gradually curve and blur. Increasingly, an imbalance develops between the

individual's capacity to understand and his semantic environment. It is difficult to argue against those who maintain that the cybernetic revolution with the ever increasing automation and computer capability, the efficient communication and identification of information, the effective means of storing and transmitting information, the easy access to enormous banks and repositories of information and data, and the artificial intelligence techniques and simulations, are quantitative rather than qualitative extensions of man's faculties. Important as such developments are, on an absolute scale they barely change the level of understood complexity or the range of application of human thought. Furthermore, automation carries with it the seeds of mediocrity. It leads to mass production of information and to experimental and computed results which are often not properly scrutinized or assessed as to their validity, meaning, and significance. Mass production, as well, leads to collectivization and the associated manipulation and control of information and data that can only be accomplished centrally by companies, organizations, or governments. In all such cases knowledge becomes a processed commodity not a cherished but freely shared acquisition. This is not the tradition of science, and is a limitation of science.

Documents can be outstandingly indexed, stored, retrieved, made available, or be conveniently used for application, but the real question remains: how can all this be transferred to the human mind, be assimilated, understood, extended, and lead to new insights and creative ideas? How can all this be compressed in generalizations that will eliminate the need of detail? How can simplicity be distilled from messy complexity? This man has to do alone. And herewith lies the limitation. Enlargement of man's intellectual capacity may increase his efficiency for conceptual thought, but would this not only be a mere postponement of the "eventual saturation of our minds?"[13] Man's imagination which propels science is endless, but man is limited, as an individual and collectively as a society. Science is not limited in answering questions, but only those questions of "the kind science *can* answer."[14]

Limits of Method, Observation, Time, Travel, and Contacts. As we discussed earlier in this book, the basic tactic of natural science is analysis: fragment a phenomenon and reduce it into its parts, analyze and study each part in isolation under controlled and well-defined conditions, and inductively derive an understanding therefrom. Measurements on or observations of processes and events occurring under uncontrolled or not well-defined conditions limit the value of measurement. But how can we isolate events that had happened billions of years ago? And how can we determine whatever changes the signal we detect might have underwent for all those years and over all that distance it traveled before reaching us? How can we establish a causal relationship between the initial events and the observation under such circumstances? Similarly, how can we infer *uniquely* the properties of the whole from the isolated and limited observation of the part knowing that the former is affected by the entirety of the system? Science is cognizant of these questions and indeed attempts to provide answers to them. For example, the effort to develop the Superconducting Super Collider (SSC) multi-billion dollar machine (it was never built) was aimed at least in part at probing the initial, transient stages of the universe by studying, paradoxically, the interactions between two of the tiniest particles in nature (two very fast protons having TeV energies).

There also exists a limit to the contacts of the scientist, especially his *direct* contacts. As the number of scientists increases so do the contacts and interactions amongst them. The scientist by stretching himself to encompass these interactions, finds himself less efficient and detached from depth. While this predicament can largely be overcome by the recent advances in and the availability of the means of easy communication such means partially solve one problem and introduce another. They often restrict the desire for and the necessity of direct personal contacts and trivialize and hinder the truly important by an unscrutinized flood of information easily exchangeable over the networks. Whether, then, the scientist's vision is limited by how deeply he goes into a particular field of science, or whether his over extension in breath limits his depth, or whether modern means of communication allow him to communicate but limit his personal contacts with his colleagues, he is always faced with the same frustrating limit: that of the self.

Metaphysical Inadequacy of Science. As we have discussed in Chapter 9, science says nothing about "the meaning of the world," "the meaning of existence," the "aims of society," "the science of the soul." Perhaps, as many have said, the "mysteries of life" were intended to remain just that: mysteries. Mysteries to be looked only from within. They, however, point to yet another limit of science: its metaphysical inadequacy. Science is thus limited by its effects (real or perceived) on metaphysical and religious ideologies, and by attempts to link science to metaphysical and even teleological questions. There are limits to the scientific world view as there are limits to the ideological views held by man. The inadequacies of both science and ideology inflict a limit to both.

10.2. LIMITS TO SCIENCE

Progressively, man's attitudes toward science (and technology) will become the most limiting factor to the functioning and support of science. There is a source of opposition to, and negative images of, science that can be characterized as extrinsic or socio-political limits to science. These restrict the appreciation of science and hinder its impact on man and his culture. They lower the stature of the scientist in society and induce a loss of confidence in the integrity and intellectual discipline of science. They fuel modern man's tendency to blame science for society's ills and to emphasize the negative aspects of technological progress while taking the positive aspects for granted. Much of the opposition to science is intrinsic to the nature of scientific knowledge, other is intrinsic to the nature of man and the increasing lack of confidence in himself, and still other has to do with modern times and the ferocity of the problems that beset the world today.

The Bureaucratization of Science. The bureaucratization of science threatens the autonomy of science, its tradition, and the conduct of scientific research. Gradually, it takes away the freedom of the scientist and puts him in the hands of appointed administrators who are often strangers to science, *ex-officio* bureaucrats who care more about appearance and public relations than essence and who often lack the necessary training, understanding, and commitment to lead the institutions where science conducts its functions. The detachment of many "managers" of scientific institutions from science and from the practicing scientists and their constant and almost total preoccupation with the "customers," the "managing contractors," the "public relations experts," the "lawyers," the "politicians," the "inspectors," the "regulators," and (even) the "social engineers," leaves no time for them to think about what the institution is all about; whatever little is left of their time, they often use to talk amongst themselves. This engenders serious threats to the proper advancement of science.[15]

The bureaucratization of science hinders the proper functioning of science, consumes and wastes its talent and resources, and weakens the integrity of science; it limits science.

The Size of the Scientific Effort. Science is not just a human activity, it is a *big* human activity. The United States 1996 spending for basic research was about $14 billion and the total 1996 federal R&D spending was about $72 billion.[16] The industrial R&D levels were even higher and since the 1980's surpassed the federal R&D by increasingly larger percentages. The bigness is a natural development of modern times, but the sheer size of the cost of modern science puts limits on science. For instance, the fact that societal agencies -- principally the government -- are responsible for the pattern of research support carries with it demands for particular results which often prevent the development of a comprehensive understanding of nature. Science conducted under such conditions risks forgoing standards, neglecting controls for disciplined experimentation, and circumventing proper disclosure of findings. Scientists and science workers are no longer dedicated solely to the truth: they are responsible for spending big money, are limited by the conditions of funding, are under pressure to keep people on the payroll, and are constantly preoccupied with the survival of their programs and institutions.

Thus, it is argued, "the advent of big science caused the corruption of the Republic of Science."[17] As with so many other of man's institutions, the sheer size of the scientific effort limits science.

The Narrow Impact of Science on the "Common" Man's Perception, Understanding, and Culture. The product of scientific research is normally the scientific paper. This product has no meaning, relevance, or value to most people. Only a small number of scientists in a restricted field normally read it; the majority of the scientists in other fields do not. Seldom a non-scientist fully recognizes the nature of scientific knowledge and its nonmaterial benefits. Neither the intricacies of relativity nor the formalisms of quantum mechanics are common sense, not even among scientists. The scientists speak of nuclear forces as described by physics, of cell division as described by biology, of molecular structures and new materials as described by chemistry, and of new regions of the cosmos as described by astronomy, but they seriously doubt their ability to speak about all of them. They are eager to direct one to the expert colleague. It is, thus, not difficult to see why the majority of society feels -- and will probably continue to feel -- an outsider to virtually all science. It seems that though the cumulative aspects of life help man understand his culture and shape his values, the cumulative aspects of science stand in his way of understanding science.

To understand and to appreciate science one has to become more-or-less an expert and this is a long and involved process. The common man's judgement, therefore, has and will probably continue to have no real validity in basic science as it does in other human vocations. Seldom for example does the non-scientist distinguish between scientific information and scientific understanding, and seldom does he understand the tentative nature of scientific findings and limitations of scientific explanations. The common man can build and operate machines and factories, but the conceptual structure of science is difficult for, if not alien to, him. Science connects to society principally through its utilitarian aspects and through its real or perceived adverse impact, not through its content or spirit.

While science has impacted modern man's culture in many and profound ways (Chapters 2 and 8), this penetration is still limited and in some instances challenged. For instance, Brooks notes that the insistence of the scientific community "on individual excellence and on vigorous interpersonal valuations runs strongly counter to contemporary egalitarian trends and rejection of all competition and comparisons between people, especially among youth."[18] And many still are suspicious of science because "What had appeared to be an instrument capable of realizing human goals turns out to set its own conditions and to impose its own values."[19] Indeed, many today not only fail to see the humanity in science,[20] but would repeat what Smyth had said almost fifty years ago, "If I had to choose between freedom and science, I would choose freedom."[21] Although Smyth went on to say that no such choice need be made because science and freedom are necessary to each other, nonetheless in the minds of a large fraction of society the implication remains that there is some kind of a choice to be made between the two.

The "two cultures"[22] are far apart, and their distrust close. In the long run, this is a severe limit to science. A limit to the cultural impact of science on society rooted in both the nature of science and the character of society.

The Expanding Utilization of Science. It is feared that the expanding utilization of science corrupts its functions and its practitioners and will thus hinder science's future. "As the power of science increases, its uses become less sacred, more trivial, more brutal, and often more immoral. Scientists are not entirely responsible for this desecration of their achievements, but we have done little to prevent it," wrote Dubois.[23] As the power of science increases, the systems it puts in motion escape man's control. The inner workings of gigantic computerized machines, for instance, are beyond the understanding by a single person. They function on the basis of rules and criteria no one knows explicitly and they are often immune to change and thus once in motion may be difficult to render inoperative.

The Fear of Science and the "Dangerous" Knowledge it Generates. As we have discussed in Chapter 2, modern man has become increasingly cognizant of the primary role science and technology play in the present and future conditions of his life, and he worries. He associates science with ecological and human perils: physics with nuclear weapons (the "ultimate" example of dangerous knowledge) and dangerous nuclear waste; chemistry with chemical warfare agents (phosgene and mustard gases), dangerous substances (thalidomide), and chemical pollutants (fluorocholocarbons and freons); mathematics with computer "data banks;" biology with biological warfare and intrusions into the sanctity life; and so on. An image of science is emerging of a force which can and does evade, destroy, threaten, and limit. A tragic polarization is developing between science and science-based technology on the one hand, and society-at-large on the other hand that limits both science and society. Interestingly, the key word, *nuclear,* was left off the name of nuclear magnetic imaging machines -- a powerful science-based[24] invention in the service of mankind -- lest the public's fear for anything "nuclear" will reject the use of the machine!

Man has reasons to be fearful of scientific knowledge for it repeatedly led to conditions he never wanted: he will for ever be under the thread of nuclear annihilation and he cannot undo this predicament. Modern man is desperately afraid of the shadow of nuclear war and the potential of science to manipulate man. He questions the ability of science to enforce "disciplined scientific procedures," and actually worries that anonymous and powerful forces may take over the control and the power of scientific knowledge. Under such a scenario, scientific knowledge is or is perceived to be perilous, a magnifier of the threat to peace and the consequences of conflict. The destructive application of science indiscriminately condemns science as producer of "dangerous knowledge," and seriously limits the advancement of science.

The Perceived Limitations Science Imposes on Man's Freedom, Privacy, and Humanity. Science is frequently painted as an ever increasing power center that limits man's freedom and privacy by enabling over control, standardization, and the take over of man's life while man remains incapable to stop these developments from happening. Science and science-based technology are accused of having a "dehumanizing effect" on society and especially its leaders. They are further accused for having replaced "individuality with mass society," and for having introduced "mechanistic philosophies" (Chapter 2). While these are broader symptoms of the lack of confidence of man in man, they are, nonetheless, connected with science and limit it.

The Challenge it Presents to Religious Beliefs, Values, and Ethics. Science, it is claimed, has removed man from the center of the universe and from his separate place among the animals. Science is accused of having set off in society the "deadly disease" of "the denial of eternal truth and absolute values."[25] Science, of course, does not do that (Chapter 9). Yet, the image of man as presented by science is perceived as been different from that of most religions. This negative image has, unfortunately, been reinforced by statements of scientists. For instance, when they claim that science denies the uniqueness of man as opposed to other animals,[26] or when they proclaim that "the cosmos (matter) is all that is or ever was or ever will be."[27] A serious problem thus arises when science becomes scientism.

Man, furthermore, blames science for moving him too fast in too many new directions that destabilize his traditional ways of life and values. He wonders aloud, for example, if the new scientific advances in life sciences are a blessing or a curse. If, he, like Crisos (the famous king of Persia long time ago), wished to be able to touch things without turning them to gold! He is confused about the ethics of *in vitro* fertilization and research with human embryos, the potential shifting of judgement of individual and social behavior from values and ethics to biology, and the extent to which a scientist's claim to *his* intellectual freedom supersedes that of society's to protect itself and its value system. This is and will continue to be a serious area of conflict and a limiting force to science.

The Destabilizing Influence Science Has on Human Institutions. We have touched on this on a number of occasions. It suffices here to summarize: Man's "stable" states are shaken by the ferocious winds of the scientific and technological hurricane of change. Established human institutions become inadequate to face the new challenges and some vanish in obsolescence. Threatening change causes uncertainty, breads insecurity, and strikes the self at its core. The loss of faith in the stable state and the loss of personal security make it difficult for individuals and for society to cope with change, and this is a limiting factor for science.

The Negative Applications of Science and its Use for Destruction. Elements of this limit have been elaborated upon elsewhere in this work (e.g., see Chapter 2). Here we summarize: There is a wide-spread negative perception of modern science by man. Many in society maintain that "man's collective passions are mainly evil," and to the devil in man, to his rivalry, and to his hate, science provides effective means of expression that threaten to destroy civilization. Science is, thus, perceived to magnify the threat to peace and this is limiting for science.

The Partnerships of Science with the Military-Industrial Complexes. "The frightening thing which we did learn during the course of the war (WWII)," said I. Rabi, "was, how easy it is to kill people when you turn your mind to it. When you turn the resources of modern science to the problem of killing people, you realize how vulnerable they really are."[28] Penicillin and polyamides are products of science, but so are nuclear weapons and napalm, pointed out a notable group of chemists.[29] Science has returned war to its primitive mode of indiscriminate killing! Worse still, it made killing easier and on grand scale! "Our country's (USA's) use of the A-bomb," wrote Bernstein,[30] "aimed largely at civilians in major Japanese cities, was part of a new strategy of killing noncombatants and terrorizing survivors into submission. This practice. .... constituted a profound transformation of values: it legitimized the mass killing of civilians as an acceptable form of war." Bernstein also made a disturbing revelation. He referred to a declassified letter from Robert Oppenheimer to Enrico Fermi according to which these scientists "did consider the possibility of poisoning half a million of the enemy with radiologically contaminated food." In the declassified 1943 letter to Fermi reporting "on the question of the radioactively poisoned foods," Oppenheimer briefly discussed the purification of beta-strontium, stressed secrecy, and proposed that "we should not attempt a plan unless we can poison food sufficient to kill a half a million men."

Will, then, the scientists continue to design ever more effective means of destruction? Most likely yes! Most of the scientists today work for the government. The fact that science and the scientist have been in the service of brutal totalitarian regimes such as those in the Former Soviet Union and Nazi Germany, regrettably gives credence to the argument of those who paint a picture of the modern scientist as a "mercenary" who has no problem in serving any master, just like Daedalus did in the ancient Greek myth three millennia ago (Chapter 2).

Science, then, is perceived as aiding totalitarianism and as irretrievably diminishing democracy by making it possible for power centers to increase their control over society. Ironically, the mastery of nature is looked upon as becoming the basis for oppression of society and repression of man. Basic research becomes the core of military power and supper weapons, and contemporary science becomes an "extension of militarism" (Chapters 2 and 6). In the eyes of the victims, the oppressed, and the rest who constantly live under the fear of modern weaponry, science and the scientist are guilty, at least guilty by association. Undoubtedly, science will continue to get a large part of its support for its use in the development of weapons and the means of destruction, but the more it does the more it will limit itself and society.

The Drafting of Science by Industry for Profit. A very large fraction of scientists today work in industry. This is a trend that will likely increase in the future as society becomes more and more dependent on technological innovation. Even universities frantically seek industrial "customers" for revenue. Basic scientific research is exploited by industry for profit in often shortsighted ways. It is feared that through its relationships with government and industry science has lost its claim as the objective arbitrator in public discussions of science-based and technology-based problems.

And science has yet another problem. It is broadly confused with technology. The distinction between the two is often not made. Perhaps the reason is that the marriage between science and technology has long been consummated and modern man recounts: *the atomic bomb came directly from the early 20th century physics, the radar from late 19th century physics, the late 20th century technology from late 20th century science, the* .... It is, however, as sad as it is disturbing to see science blamed for every industrial disaster and industry to be credited by every technological success! The incestuous relationships of science, technology, and industry diminished society's faith that science is a beacon of openness and the scientist is loyal to the ideals of society and truth. History is replete of instances where decisions on *how* technology is to be used in society were insensitive to the needs of those affected by it (Chapters 2 and 6). The recent advances in genetics research and biotechnology have led to renewed negativism. Scientists are said to becoming gene merchants, and industrial partners in the genomics gamble. A number of recent reports in *Science* magazine are replete with statements highlighting the seriousness of this issue. "Drug companies, biotechs, and Wall Street investors are putting their money down on efforts to unlock the secrets of human DNA," wrote *Science* in one of its recent issues.[31] The same magazine announced a year later, "A nonconformist sequencer (geneticist J. Craig Venter) has teamed up with a company in a crash project to sequence the human genome."[32] It is feared that genetics research may be restricted by undue regulation and anti-genetics referenda.

False Promises and Excessive Expectations. Society largely pays for modern science. Society, thus, expects science to deliver on its investment. There are serious problems as to what science is expected to do for society at any given time and both science and society share the blame. There is perhaps a limit as to the perceived relation of knowledge to relevance. A limit to what science can promise and a limit to what society can demand, which is not fully recognized. While more scientific knowledge can be expected to effect more relevance, science cannot be expected to cure all ills of society. Since WWII society has come to believe, especially in the United States, that science can solve any problem of society if only appropriate levels of support were provided; if knowledge or materials are not available for the solution, science will come up with adequate know-how to get the job done. Real confidence, possibly too much confidence. Confidence nonetheless that has been strengthen time-after-time by developments since WWII. In fact, even possible-in-principle big undertakings are deliberately pushed forward simply because the sheer pursue of a science-based technological goal which at a given time lies beyond the full capability of science can serve as a central force for projecting beyond the state-of-the-art into new areas of knowledge, techniques, materials, and technology, and into new discoveries and innovations. Indeed, it has been argued that even the Strategic Defense Initiative (SDI) program in the United States was worth undertaking for just this purpose: as a focused search effort for new knowledge and materials for the next level of technology, military and civilian. This, then, carries the danger of false expectations and limits science.

The Exploitation of Science by Politics and the Influence of Politics on Science. In this century, but especially since WWII, politics has learned to explore and exploit science and science has become more and more political. Governments enlisted their scientists for nationalistic goals during war (particularly during WWI and WWII) and peace, and scientists eagerly served their politicians in various roles and missions, allowed themselves and science to be manipulated and exploited by the politicians and the executive branches of government, and profited in the process. Initially it was mostly physicists, then chemists, now is the biologists' turn. Initially they were few, mostly basic scientists, now they are many, mostly of "general" backgrounds.

The politicalization in science[33] allowed politicians and governments to rationalize, justify, and legitimize economic and political actions and policies often contrary to the wishes of the people. It immersed the scientist in national and political struggles in which he is often perceived to have circumvented the norms of science and to have used the prestige of science to wield power and profit. He has often crossed the threshold that separates science from trans-science (Section 10.3), and dealt with trans-scientific questions in scientific terms without distinguishing between his opinion and the facts of science. He, regretted some of his actions, like J. Robert Oppenheimer at Alamogordo, but *only after the fact.* Indeed, many wrote on this issue, but it suffices here to quote the disturbing conclusion reached by Haberer.[34] "What becomes evident is that scientific leaders, when faced with a choice between the imperatives of conscience and power, nationalism and internationalism, and justice and patriotism, invariably gravitated toward power, nationalism, and patriotism, and followed a policy of prudential acquiescence." Indeed, politicalization and nationalism limit science's freedom and impartiality.

There is no way science will escape its involvement with politics and there is, thus, no way science will escape the criticism that involvement engenders. Criticism will thus continue to be levied against science, and dissenting voices will continue to be heard, such as those over three decades ago calling for science to be placed "under more strict control because with its present freedom it represents an imperialistic imposition on modern society and an autonomous sphere with excessive political power,"[30] These attitudes will limit science.

The Errors of Scientists. As persons in other walks of life and professions, scientists showed ineptness and committed errors (Chapter 6): when they claim excessive intellectual right of freedom, when they place the satisfaction of their curiosity above the needs of society, when they are part of scientific and technological programs in which they bend their ways and deviate from disciplined scientific procedures, when they give unsubstantiated advice, when they promise what they cannot deliver, when they deal with trans-scientific questions in scientific terms, or when they use the press to promote their work and sell their science. For better or for worse the scientist, like the priest, has been held to a higher standard than persons in most other vocations. Unquestionably, the stature of the scientist has been lowered recently and the confidence of society in him has declined. This may be a symptom of a more general decline in the confidence of man in his fellowman, but it is no less discomforting or any less limiting.

Regulated or Forbidden Science. The issues we discussed in this chapter individually and collectively refer to the constrains on science by the laws of nature, the state-of-the-art technology and method available at any given time, and the structure, values, and laws of society. Especially with regard to societal type constrains, the question "What should science do and how should science go about doing it?" is accompanied by a series of other questions which are answered differently within and outside science. Questions such as: Should aspects of science be classified as "forbidden," and should certain kinds of knowledge be classified as "inappropriate?" Could there be knowledge the possession of which is harmful to society and thus further accumulation of knowledge inappropriate? Should constraints be imposed and if so at what level should they be imposed and who should impose and administer them? Is societal control of science (and technology) possible? Is it inevitable? How valid is the argument that if we do not do a particular type of research others in another laboratory, or in another country, or at another time would? Does the scientist have the right to do what society does not want him to do, if he himself or a private organization pays for it? Is the primacy of unhindered right to new knowledge absolute, or does it have to be moderated by legitimate concerns of society?

Whatever the answers to these questions may be, it is certain that man, as scientist and as citizen, will be limited by the burden of responsibility the power of scientific knowledge imposes on him. The limits must be commensurate with the power if conflict and remorse are to be avoided.

10.3. TRANS-SCIENCE

So often society is faced with big issues which go beyond the possibility of exact scientific resolution. Issues which lie at the interfaces of science and politics and science and society. Questions that can be stated in scientific terms, but cannot be answered by science. Questions that seem to be a part of science, but in fact transcend science. Questions incapable of resolution by science because at the time they are posed they are in principle "beyond the proficiency of science."[35] For such questions A. Weinberg[35] proposed the term *trans-scientific*. He suggested that the health effects or risks associated with low-level environmental insults (ionizing radiation and chemicals) fall in this category. Low-level effects are buried in the noise, if they exist, and require a large number of tests and observations to demonstrate an unequivocal effect. Estimates of extremely unlikely events cannot be made with the scientific validity applied to estimates of events for which there are abundant statistics. An example of a current trans-scientific question is the issue of possible health effects from magnetic fields generated in the transmission and distribution of electric power.

Trans-scientific questions will always be around and they will be of political and social significance because they enter the assessment of risk versus benefit process for which the available information is often limited.[36] Answers to such questions are sought because decisions have to be made, such as the enactment of a law or a regulation. Trans-scientific questions require wisdom along with scientific competence and they may have to be looked at -- and be followed in time -- from different perspectives and from the stand points of different scientific fields. "What the scientist can do in clarifying matters of trans-science differs from what he can do in clarifying matters of science. In the latter case, he can bring to bear his scientific expertise to help establish scientific truth. In the former case, he can at most, help delineate where science ends and trans-science begins," instructed Weinberg.[37] The interfaces of science with politics and decision making represents a form of limit to science, for if one trans-scientific question is ultimately resolved another is likely to appear.

There are perhaps as well interfaces of science with other human activities that can be similarly characterized as trans-scientific. Areas that lie at the interface between science and religion, science and philosophy, science and "meta-science." Weinberg's terminology may, thus, be expanded to include questions in these other interfaces of science. Questions which can be stated scientifically, but which -- unlike the questions for which Weinberg coined the word trans-scientific which at some future time may be answered by science -- transcend science possibly for ever because they lie outside the scientific domain now and possibly in the future.

Two examples may suffice to illustrate this type of trans-scientific questions. The first example is the question regarding *the origin of life*. We can define the problem scientifically. Life exists. It, therefore, must have a beginning. Thus the question: *What is the origin of life?* Here is a sample of how scientists recently responded to this question.[38]

- N. Mott: "It arose through some chemical reaction in the primeval mud."[39]
- R. Jastrow: "Nobody has demonstrated that life, even a simple bacterium, can evolve from a broth of molecules."[40]
- R. A. Naumann: "Given the right precursors and conditions, living systems will arise spontaneously."[41]
- C. H. Townes: "I do not know how life originated."[42]
- I. Prigogine: "Life is a cosmic phenomenon, which appears when the right conditions are satisfied."[43]
- W. Arber: "Life only starts at the level of a functional cell .... How such already quite complex structures may have come together, remains a mystery to me."[44]
- H. Uhlig: "A miracle."[45]
- E. P. Wigner: "The origin of life is a mystery."[46]
- A. L. Schawlow: "Even if the origin of life can eventually be broken down to a series of chemical steps, subject to known physical laws, it will still be marvelous that those powerful laws have such enormous potential."[47]

Scientists have views about the origin of life, but they have no scientific answer to the question. The question is trans-scientific. It lies beyond today's science.

The second example: *What is the origin of the universe?* This question was again chosen because of the recent views of a wide spectrum of scientists on the subject.[38]

- I. Prigogine: "The universe started from an instability of the quantum vacuum."[43]
- R. Merle d' Aubigne: "My mind ....is incapable of conceiving the Nothing,

before or after the existence of the universe."[48]

- R. Jastrow: "The fact that the universe was once in a dense, hot state is considered proven by almost every scientist .... What forces filled the universe with energy fifteen billion years ago?"[40]
- W. A. Little: "I go along with the Big Bang picture but recognize that it does not address the deeper issue as to why it happened."[49]
-V. Kistiakowsky: "There remains the question of how the Big Bang was initiated, but it seems unlikely that science will be able to elucidate this."[50]
- E. Nelson: "It is premature to investigate the origin of the universe."[51]
- E. Segré: "The origin of the universe, at present, does not seem to me to be a scientific question. Scientific theories are usually validated by experiment, consistency tests, and predictive power, all of which are hardly applicable to the origin of the universe."[52]
- L. Neel: "As a physicist, I consider physics to be an experimental science. A hypothesis is of interest only if it is possible to verify its consequences by discovering new phenomena or new directions. This means that all hypotheses concerning the origin of the universe do not belong to physics but to metaphysics or to philosophy and that physicists as such are not qualified to deal with them."[53]
- U. J. Becker: "Scientifically unknown."[54]
- E. P. Wigner: "The origin of the universe is a mystery for science, surely for the present one."[46]
- S. Bowyer: "Ultimately, the origin of the universe is, and always will be, a mystery."[55]
- H. Margenau: "God created the universe. .... and the laws of nature."[56]

Views on the question, but no answers! Questions rather than answers! Questions not any simpler than that of Leibnitz years ago: "Why is there something rather than nothing?"[57] And yet living systems and the cosmos do exist and science poses the question: where did they come from and how? A trans-scientific question indeed that exemplifies science's limits.

10.4. REFERENCES AND NOTES

1. E. P. Wigner, Proceedings of the American Philosophical Society **94**, 422 (1950); P. Auger, Bulletin of Atomic Scientists **XXI**, November 1965, p. 21; Daedalus, *Limits of Scientific Inquiry*, Spring 1978; P. Medawar, *The Limits of Science*, Oxford University Press, Oxford, 1984.
2. M. Polanyi, *Personal Knowledge,* The University of Chicago Press, Chicago, Illinois,

1958.

3. G. S. Stent, Science **187**, 21 March 1975, p. 1052.

4. C. N. Martin, *The Role of Perception in Science*, Hutchinson of London, 1963, p. 16.

5. B. Glass, Science **171**, 8 January 1971, p. 23.

6. B. Blumberg, in P. Coveney and R. Highfield, *Frontiers of Complexity*, Fawcett Columbine, New York, 1995, p. xiii.

7. V. F. Weisskopf, Science **176**, 14 April 1972, p. 138.

8. See, for example, E. P. Wigner, Foundations of Physics **1**, 35 (1970); N. Bohr, in *Light and Life*, W. D. McElroy and B. Glass (Eds.), The Johns Hopkins Press, Baltimore, 1961, p.1.

9. E. P. Wigner, Foundations of Physics **1**, 35 (1970).

10. P. W. Anderson, Physics Today, July 1991, p. 9; P. Coveney and R. Highfield, *Frontiers of Complexity*, Faucett Columbine, New York, 1995; R. Osman, K. Miaskiewicz, and H. Weinstein, in *Physical and Chemical Mechanisms in Molecular Radiation Biology*, W. A. Glass and M. N. Varma (Eds.), Basic Life Sciences, Vol. 58, Plenum Press, New York, 1991, p. 423; Science **284**, 2 April 1999, pp. 79-109.

11. C. S. Smith, in *On Aesthetics in Science*, J. Wechsler (Ed.), The M.I.T. Press, Cambridge, MA, 1979, p. 22.

12. P. Medawar, *The Limits of Science*, Oxford University Press, Oxford, 1984, pp. 68-82.

13. P. Auger, Bulletin of Atomic Scientists **XXI**, November 1965, p. 21.

14. Ref. 12, p. 87.

15. Many authors have discussed this topic in both general terms (e.g., J. Ziman, *Prometheus Bound*, Cambridge University Press, Cambridge, 1994.) and in specific terms (e.g., R. P. Crease and N. P. Samios, The Atlantic Monthly, January 1991, p. 80).

16. P. V. Dominici, Science **273**, 6 September, 1996, p. 1319; N. F. Lane, in *Atomic and Molecular Data and Their Applications*, P. J. Mohr and W. L. Wiese (Eds.), AIP Conference Proceedings 434, American Institute of Physics, Woodbury, New York, 1998, pp. 3-6.

17. M. Polanyi, Minerva **I**, Autumn 1962, p. 54; Minerva **V**, Summer 1967, p. 533.

18. H. Brooks, Science **174**, 1 October 1971, p. 21.

19. D. A. Schon, *Beyond the Stable State*, Random House, New York, 1971, p. 22.

20. J. Goodfield, Science **198**, 11 November 1977, p. 580.

21. H. D. Smyth, American Scientist XXXVIII, July 1950, p. 426.

22. C. P. Snow, *The Two Cultures: and A Second Look*, Cambridge University Press, Cambridge, 1965.

23. R. Dubois, *Reason Awake*, Columbia University Press, New York, 1970, p. x.

24. F. W. Wehrli, Physics Today, June 1992, p. 34.

25. Jacques Maritain, quoted in G. Holton, Science **131**, April 1960, p. 1188.

26. J. Bronowski, *The Identity of Man*, The Natural History Press, Garden City, New York, 1965, Chapter 1.

27. C. Sagan, *Cosmos*, Random House, New York, 1980, p. 4.

28. I. I. Rabi, *Science: The Center of Culture*, The World Publishing Co., New York, 1970, p. 71.

29. F. A. Long, R. S. Morison, F. H. Westheimer, A. M. Bueche, and D. F. Horning, Chemical and Engineering News, August 25, 1969, p. 52.

30. B. J. Bernstein, Technology Review, May/June, 1985, p. 14.

31. J. Cohen, Science **275**, 7 February 1997, p. 767.

32. E. Marshall and E. Pennisi, Science **280**, 15 May 1998, p. 994.
33. F. von Hippel and J. Primack, Science **177**, 29 September 1972, p. 1166; K.-H. Barth, Physics Today, March 1998, p. 34.
34. J. Haberer, Science **178**, 17 November 1972, p. 713.
35. A. M. Weinberg, Minerva **10**, April 1972, p. 209.
36. G. Majone, *Science, Technology, and Human Values* **9**, Winter 1984, p. 15.
37. A. M. Weinberg, Science **177**, 21 July 1972, p. 211.
38. H. Margenau and R. A. Varghese, *Cosmos, Bios, Theos,* Open Court, La Salle, Illinois, 1992.
39. N. Mott, in Ref. 38, p. 64.
40. C. R. Jastow, in Ref. 38, p. 45.
41. R. A. Naumann, in Ref. 38, p.70.
42. C. H. Townes, in Ref. 38, p. 123.
43. I. Prigogine, in Ref. 38, p. 188.
44. W. Arber, in Ref. 38, p. 141.
45. H. Uhlig, in Ref. 38, p. 125.
46. E. P. Wigner, in Ref. 38, p. 131.
47. A. L. Schawlow, in Ref. 38, p. 105.
48. R. Merle d' Aubigne, Ref. 38, p. 157.
49. W. A. Little, in Ref. 38, p. 54.
50. V. Kistiakowsky, in Ref. 38, p. 51.
51. E. Nelson, in Ref. 38, pp. 75 and 76.
52. E. Segré, in Ref. 38, p. 109.
53. L. Neel, in Ref. 38, p. 73.
54. U. J. Becker, in Ref. 38, p. 28.
55. S. Bowyer, in Ref. 38, p. 31.
56. H. Margenau, in Ref. 38, p. 57.
57. R. Jastrow, in Ref. 38, p. 49.

CHAPTER 11

The Future of and in Science

Through the centuries man's opinion about the future converged on at least two aspects: that the future is unknown and that it must be hopeful to be desired. Modern man can make no claim of predicting the future, but he is free, as his predecessors, to hope and to dream. Free to have a distant and perhaps detached view of the unknown. Free to judge the size of the ocean and the beauty beyond not merely by what his eye can see, but by the grasp of his dream and the excitement and promise it entails. Emotion, the bonds of history, the constraints of culture, and the conditioning of knowing limit one's vision, and distance blurs the nature of reality. Yet, of what good is the short-sighted vision? Like Thomas Jefferson, I, too, "like the dreams of the future better than the history of the past,"[1] but I also like the beacons of inextinguishable hope of the past, the Phoenix of the Greeks and the Easter of Christianity. Symbols of hope based on sacrifice and love of humanity.

Neither the social planner, nor the social engineer, nor he who thinks "that the future will follow exactly the past" can make the dreams of a longing humanity come true. This privilege belongs to him who dreams of a hopeful future where everything that can go well will go well and where everything that may go wrong can and will be prevented. A prerequisite of this fundamental hope is the knowledge that science will provide for him and the enlightenment with which he will use it.

11.1. SCIENCE AHEAD

Unpredictability is the only predictable element of science's future. Virtually none of what we think will happen, will happen the way we thought it would. Indeed, we can predict with certainty that the most significant things that will happen in science will be the ones we had never predicted. This is the beauty of adventure. Nonetheless, one can foresee certain trends in science in the near future and below we point out a few.

Science Will Continue its Expansion Outward and Inward. Outward toward the very old and the very big and inward toward the very small and very short lived. In this, science will develop new techniques and will build bigger and more expensive machines both for space (space probes, space stations, telescopes, missions to planets) and for here on earth (reactors, accelerators, light sources). It will be exploring neutron stars, pulsars, quasars, black holes, and other celestial bodies yet undiscovered; it will be searching at the outer fringes of the cosmos; it will be looking for extraterrestrial life; it will be pondering over the origin and the future of the cosmos; and it will be theorizing anew about it all. Science will be also digging deeper into the interior of the "cosmic onion," inside the atomic nucleus and elementary particles, searching for new insights about the nature of forces, the fundamental laws of matter, the interactions between the basic units of matter and antimatter, and the modularity of nature even at the level of the very small.

Science Will Expand Toward Complexity and Toward the Intricate Manipulation of Nature. Both of these trends will push science -- foremost physics and chemistry -- toward the understanding of broader areas of nature and the use of this understanding to benefit man. The study of atoms, molecules, ions, and radicals in isolation and within varied environments will reach an ever increasing level of complexity and sophistication, especially in conjunction with the manipulation of the environment to effect hitherto novel properties of materials and surfaces. There will be exploration and exploitation of the interplay of radiation and matter and charge and light; the interactions of and with energy-rich species; the properties of matter under extremely high electric fields, extremely low and extremely high temperatures and pressures; special properties of all phases of matter (solid, liquid, gas, plasma) and the interfaces between the phases; systems far from equilibrium; amorphous materials; surfaces; control of plasmas; high temperature superconductors; new microscopies and atomic engineering techniques employing particles and laser radiation; manipulation of molecules and modification of molecular structure for whatever the need or purpose; new compositions of biosubstances; substitute design chemicals and pharmaceuticals; biometrology.

Most profoundly, the fields of basic science -- especially physics and chemistry -- will drift toward and stimulate advances in biology and medicine. There will be an increase in the flow of the instrumentation, the method, and the content of physics and chemistry (and also of mathematics and computer science) to life sciences and biomedicine. Such instrumentation will surely include electron microscopes, electron, ion and radiation sources, lasers, magnetic resonance devices, and even big machines such as nuclear reactors for neutrons and accelerators for energetic ions to unravel the complexity of biological structures and fingerprint their reaction mechanisms with radiation and chemicals. Methods for handling large molecules in vacuum (such that one knows exactly where they are), fundamental reactions down to the electron-volt level where physics and chemistry meet most of biology, radiobiological and radiochemical efforts attempting to uncover the ways biology traces back to chemistry and physics, and attempts to describe life at the physicochemical level. Science will be moving toward the complexity of living matter and toward achieving control over biological processes -- in much the same way physics and chemistry had achieved control over physical and chemical processes -- and with that an understanding of the molecular basis and, consequently, control of disease.

Science Will Embrace the Earth. Science will explore and will attempt to better understand the earth: the terrestrial environment, the energy sources and resources of the earth, the inaccessible regions of our planet such as those covered by ice and water, the upper layers of the atmosphere, earth's tectonics and interior, and its destructive capacity (its earthquakes, its weather, its monsoons). Science will be used to protect man and his environment including his cultures.

Science Will Continue to Feed Technology and Underpin its Advances. Through science man will find new ways to manage, use, conserve, and extend his energy sources and formulate new approaches to solve the associated issues of pollution, radioactive waste, and energy storage. Man's growing electric environment will undoubtedly see many science-based innovations especially in the way and in the form electric energy is delivered and used. Advances in understanding the immensely complex biological systems will further fuel the revolution in medicine and health care systems whether in delivering new chemicals, or in using radiation and isotopes for biomedical research, diagnostic and therapeutic medicine, or in understanding the nervous system, brain, or memory. In all likelihood, the diverse technological impact of chemical synthesis on man will stretch to the new frontiers of molecular electronics,[2] and imitations of life.[3] Similarly, science will continue unabated the revolution in information technology.

At its Foremost Frontiers, Science Will Keep Encountering Religion and Philosophy. Science will continue to be bewildered by singularities, the mysteries of the beginning of the cosmos, the appearance of humanity, the absoluteness of the speed of light. It may even venture extension into the regions of consciousness and the non-traditional functions of scientific inquiry.

Thus, in spite of the unpredictability of science's future, we can predict with certainty that knowledge will go forward and science will continue to be man's endless frontier!

11.2. SCIENCE, SOCIETY, AND THE SCIENTIST: THE NEW FAIR

Science will become an increasingly larger part of the social and political process, and will come closer to man principally through its technological impact and the changing conditions of life it will effect. It will present man not only complementary pictures of reality, but also complementary approaches to the problems of life. To be successful it will have to preserve its integrity, continue its highest allegiance to the truth as defined by the validation procedures of the scientific process, and serve all societies and all peoples. Science will complete the integration of the world and will help unify man's microcultures and civilization. *A strong science will be a necessary, but not a sufficient condition for a better society.*

Society will rely more on science for its problems and needs and will experience yet more amplification of human power and thus increased potential for good and evil as a result of scientific advances. The central problem of the future will be the ethical issues relating to the control of man over his own biological evolution and the concentration of power and its destructive consequences. With the aid of science, society will have the opportunity to humanize the earth and for this it must *keep man dignified.*

The scientist must bring science and society closer together and must make science a larger part of man's intellectual tradition. This will require more breath, wider perspective, and an open mind of hope. Along with his responsibility to the truth he must uphold his responsibility to man, for he will face bigger private struggles with his consciousness and inevitably and regrettably he will know sin, again. Social conditioning or moral choice should never choke the youthfulness of the spirit that qualifies him to be a scientist. He should struggle to be a scientist and to deserve the honor!

11.3. REFERENCES AND NOTES

1. This phrase, "I like the dreams of the future better than the history of the past," was written by Thomas Jefferson in a letter he wrote in 1816 to John Adams (C. P. Haskins, American Scientist **58**, January/February 1970, p. 23).
2. As a 1979 National Academy of Sciences report (*Science and Technology,* National Academy of Sciences, W. H. Freeman & Co., San Francisco, 1979, p. 177) explains, "The ultimate goal here would be to synthesize molecules that would act as individual circuit elements, such as conductors, resistors, capacitors, etc., and then to combine these elements into amplifiers, memory devices, and so on."
3. For instance mimicking chemical processes in living organisms such as in synthetic rubber and atmospheric nitrogen fixation.

Appendices

Appendix A. Prefixes

Prefix	Symbol	Decimal point
		to the *left* by
deci	d	1 place, 10^{-1}
centi	c	2 places, 10^{-2}
milli	m	3 places, 10^{-3}
micro	μ	6 places, 10^{-6}
nano	n	9 places, 10^{-9}
pico	p	12 places, 10^{-12}
femto	f	15 places, 10^{-15}
atto	a	18 places, 10^{-18}

Appendix A. Prefixes (continued)

		to the *right* by
deca	da	1 place, 10^1
hecto	h	2 places, 10^2
kilo	k	3 places, 10^3
mega	M	6 places, 10^6
giga	G	9 places, 10^9
tera	T	12 places, 10^{12}
peta	P	15 places, 10^{15}
exa	E	18 places, 10^{18}

Appendix B: Fundamental Constants[a,b,c,d]

Constant	Symbol	Value
Speed of light in vacuum	c	2.997925×10^8 m s^{-1}
Electronic (elemental) charge	e	1.602177×10^{-19} C
Electron rest mass	m	9.109389×10^{-31} kg
Proton rest mass[e]	M_p	1.672623×10^{-27} kg
Neutron rest mass	M_n	1.674928×10^{-27} kg
Planck's constant	h	6.626075×10^{-34} J s
Boltzmann's constant	k	1.380658×10^{-23} J K^{-1}
Molar gas constant	R	8.314510 J mol^{-1} K^{-1}
Gravitational constant	G	6.67259×10^{-11} m^{3}kg^{-1}s^{-2}
Avogadro's number	N_A	6.022136×10^{23} mol^{-1}

[a] To six decimal figures.
[b] The values listed are from E. R. Cohen and B. N. Taylor, Physics Today, August 1990, p. 9. See this reference for values of other fundamental physical constants.
[c] Mass-energy conversion factors:
   1 atomic mass unit (1 u) = 1.660540×10^{-27} kg = 931.49432 MeV/c^2
   1 electron mass = 0.510999 MeV/c^2
   1 proton mass = 938.27231 MeV/c^2
   1 neutron mass = 939.565632 MeV/c^2
[d] Quantum energy conversion factors:
   1 eV = 1.602177×10^{-19} J = 1.602177×10^{-12} erg = 8065.75 cm^{-1} (4.18×10^3 J = 1kcal)
   Wavelength λ of photon with energy E (in eV): λ(Å) = λ(nm x 10) = $(12397.67)/E$ (eV).
[e] Ratio of proton to electron mass: $M_p/m = 1836.152701$.

Appendix C: Two Examples from Little Science Illustrating the First Stretch that Leads from Basic Research to Application

In this Appendix the translation of basic scientific results to technological products is illustrated by two examples from low-energy electron collision physics. Let us then first elaborate on the unique reactions of slow electrons with matter, and for our purpose here, with molecules. Slow electrons we call those with kinetic energy, ε, lower than about 100 eV.[1] They are one of the most fundamental, abundant, and reactive particles in nature. They are generated by a multitude of mechanisms in all states of matter.[2,3] Once generated in a medium -- say, in a gas made up of molecules -- the slow electrons interact with the gas molecules in a number of unique ways. Some of the most significant of these interactions are those in which electrons lose energy via elastic and inelastic collisions with the gas molecules symbolized by the reactions

$$
\begin{aligned}
e\,(\varepsilon)\;+\;AX\;&\to\;AX\;+\;e\,(\varepsilon) & &\text{elastic electron scattering} & &(1a)\\
&\to\;AX^{*}\;+\;e\,(\varepsilon') & &\text{inelastic electron scattering} & &(1b)\\
&\to\;A\;+\;X\;+\;e\,(\varepsilon') & &\text{molecular dissociation.} & &(1c)
\end{aligned}
$$

In the above reactions AX represents an unexcited molecule, AX^{*} an excited molecule, $e(\varepsilon)$ an electron with kinetic energy ε before the collision, $e\,(\varepsilon')$ an electron with kinetic energy $\varepsilon' < \varepsilon$ after the collision, and A and X represent atomic or molecular fragments with or without excess internal energy. Another group of significant processes via which electrons interact with molecules are the "ionizing" collisions. These are collisions in which additional free electrons are generated in the gas when the colliding electron has sufficient energy to ionize the gas molecule. They can be represented by the reactions

$$
\begin{aligned}
e\,(\varepsilon)\;+\;AX\;&\to\;AX^{+}\;+\;2\,e & &\text{ionization} & &(2a)\\
&\to\;A\;+\;X^{+}\;+\;2\,e & &\text{dissociative ionization.} & &(2b)
\end{aligned}
$$

There is a third type of reactions which is characteristic of electrons with energies below about 15 eV, namely, indirect (capturing) collisions. These are collisions in which the slow electron enters an empty orbital of the molecule and occupies it for a while becoming temporarily captured by the molecule. The negative ion formed this way is temporary, transient, because it has excess energy and is unstable toward electron ejection. Its lifetime (the time the captured electron stays on the molecule) can be as short as few femtoseconds and as long or longer than milliseconds.[2,3] However, when the transient anion is autodestructed by electron ejection, the neutral molecule may retain part or all of the electron's initial energy becoming itself excited. Alternatively, the transient anion AX^{-*} can be destructed (decay) by autodissociation forming permanent fragment negative ions X^- and free radicals. The autodissociation process takes place when it is energetically possible and when the fragment X (and/or A) has a positive electron affinity (that is, when X can "permanently" bind the extra electron). In addition, the transient anions AX^{-*} of many molecules with positive electron affinities can rid themselves of the excess energy (symbolized by the asterisk, *) mostly by collision with other molecules and form stable parent AX^- ions. Pictorially, these processes can be written in the following way

$$e\,(\varepsilon) + AX \rightarrow AX^{-*} \rightarrow AX^{(-*)} + e\,(\varepsilon') \qquad\qquad (3a)$$
$$\rightarrow A + X^- \qquad\qquad (3b)$$
$$\rightarrow AX^- + \text{energy}. \qquad\qquad (3c)$$

Such indirect collisions occur only when the slow electron has energies in restricted energy ranges in the neighborhood of the energies of the empty orbitals which the electron occupies when captured. These energy regions define the positions of the so-called negative ion states (or resonances) of molecules.

Negative ion states occur abundantly in nature. Their energy positions and numbers are distinctly characteristic of the electronic structure of molecules. There are molecules (e.g., N_2) and atoms (e.g., rare gases) which exhibit only Reaction (3a) (these are called non-electronegative because they do not form stable negative ions, they have a negative electron affinity). There are molecules which exhibit Reactions (3a) and (3b) (e.g., CF_4,) (the negative ion states of these molecules lie energetically high enough to allow the transient anion to dissociate into fragments which can bind an extra electron). There are other molecules (e.g., SF_6) which exhibit all three reactions, and still others (e.g., O_2) which exhibit Reactions (3a) and (3b) in one energy range and Reactions (3a) and (3c) in another.

There are of course other particles (e.g., photons, ions) besides electrons whose interactions with the gas molecules are important. However, our knowledge of the ways slow electrons interact with molecules just described [Reactions (1) to (3)], allows control in electrically-stressed matter (matter under an applied electric field such as in a gas discharge) of (i) the kinetic energies and transport properties of electrons, and (ii) the electron production and loss, that is, the number of electrons in the gas and the conversion of electrons to negative ions and vice versa.[4] These capabilities are basic to many advanced technologies dealing with laser development, radiation detectors, dielectric materials, plasma-based technologies, particle beams, analytical instruments, environmental pollutant detection, surface modification, plasma etching and deposition in semiconductor industries, and the tailoring of materials with a variety of new properties.

Let us now see how such knowledge was translated to two advanced technological products: *efficient CO_2 lasers, and gaseous insulators for the transmission and distribution of electric power.*

Development of an Efficient CO_2 Laser. Lasing action in pure CO_2 gas was discovered[5] in 1964. The CO_2 laser radiation originates from transitions between the vibrational levels of the ground state of the CO_2 molecule. The relevant energy-level diagram of the CO_2 molecule is shown on the left-hand-side of Figure A.1. The lasing transitions are those between the 00^01 and the 10^00 (10.4 μm) and the 02^00 (9.4 μm) vibrational levels.[6] The lasing state may be populated by direct electron-CO_2 collisions and this can be done by controlling the density, N, of the CO_2 gas and the applied electric field, E, in such a way that most of the electrons that are present in the electrically stressed gas have energies where the excitation cross section for the lasing state is maximum. However, this process is inefficient.

Basic studies on electron scattering from the CO_2 molecule have shown that the cross sections for excitation to the non-lasing and lower states are also significant in comparison to the 00^01 state cross section and this causes the low efficiency observed in the laser using pure CO_2. *How, then, can the efficiency of the CO_2 laser be improved?* By pumping the lasing state more efficiently of course. Is there a way to do this? To answer this question let us first look at the right hand-side in Figure A. 1. It is observed that the first vibrational level, $v = 1$, of the ground-state N_2 molecule is isoenergetic with the lasing level 00^01 of the CO_2 molecule. If, then, a CO_2/N_2 mixture were to be used as the lasing medium instead of pure CO_2, and if it were possible to find a way to produce efficiently vibrationally excited N_2 molecules in the $v = 1$ level, $N_2^*(v = 1)$, it would be possible to pump efficiently the lasing state 00^01 in CO_2 via the process

$$N_2^*(v = 1) + CO_2\,(00^00) \rightarrow N_2\,(v = 0) + CO_2^*\,(00^01). \qquad (4)$$

That is, it would be possible to generate excited CO_2^* (00^01) via energy transfer from $N_2^*(v = 1)$. The efficiency of this process is high because the excitation energy of the $v = 1$ level of N_2 is almost equal to that of the lasing state, 00^01, of CO_2. How, then, can we efficiently produce vibrationally excited $N_2^*(v = 1)$ in a lasing CO_2/N_2 mixture? Reaction (3a) holds the answer.

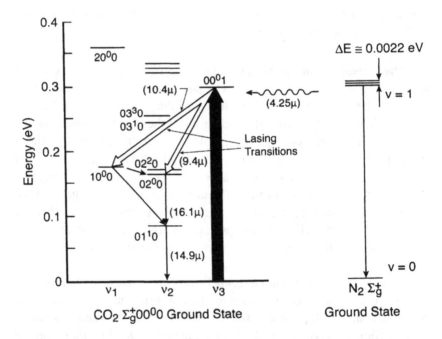

Fig. A. 1. Simplified energy-level diagram for the CO_2 and N_2 molecules illustrating the lasing transitions in the CO_2 laser (from De Maria[7]).

Basic research has shown that Reaction (3a) is an effective way to *excite vibrations of molecules.* More specifically, basic research has quantitatively measured the cross section for excitation of the N_2 molecule into the $v = 1$ state by electrons indirectly scattered via the lowest negative ion state of N_2 at 2.3 eV, that is, via the reaction

$$e\ (2.3\ eV) + N_2 \rightarrow N_2^{-*}\ (5 \times 10^{-15}\ s) \rightarrow N_2^*\ (v = 1) + e\ (slow). \qquad (5)$$

Electrons with energies around 2.3 eV are temporarily captured by the N_2 molecule forming the lowest negative ion state, N_2^{-*}, of N_2. When the N_2^{-*} transient negative ion decays by autodetachment within femtoseconds of its formation most of the electrons are released with the same energy they had when they were captured, but an appreciable number are released with less energy leaving the N_2 molecule vibrationally excited (mostly in the first vibrational level $v = 1$). Thus, Reaction (5) is an efficient means to employ to produce vibrationally excited nitrogen molecules $N_2^*(v = 1)$. This is clearly seen by the large enhancement around 2.3 eV in the cross section for electron scattering from N_2 shown in Figure A. 2 due to this process.

Here, then, comes the first step in the application process. The desirable vibrationally excited N_2 molecules, $N_2^*(v = 1)$, can be generated efficiently in a CO_2/N_2 mixture via the decay of the transient negative ion N_2^{-*} if the ratio E/N is arranged so that most of the electrons present in the electrically stressed gas have kinetic energies around 2.3 eV to produce the N_2^{-*} transient species. This can easily be done and the indirectly produced vibrationally excited $N_2^*(v = 1)$ molecules act as an excitation source for the 00^01 lasing state of CO_2. In this way the efficiency of the CO_2 laser has been significantly increased.

The percentage composition of the mixture and the addition of other gases to allow the laser to function in a continuous mode were the task of applied research and development. They were guided by knowledge on the lifetimes and collision quenching rates of the various states involved, as well as by Boltzmann-code analyses to optimize the discharge parameters and compositions which themselves required further knowledge from basic research on the collision cross sections and transport coefficients of the component gases. An extremely useful laser for basic and applied research and for industrial and medical technologies has been developed with the help of pure science. The CO_2 laser *is*, really, a CO_2/N_2 laser.

Gaseous Insulators for the Transmission and Distribution of Electric Power. This example is closely related to the knowledge just discussed but extended a little further. Through Reactions (3b) and/or (3c) are provided effective means to remove electrons from a given medium by transforming them from free and highly mobile particles to slow moving bulky negative ions. Figure A. 3 shows[9] a sample of the findings of basic science regarding reactions (3b) and (3c). A most distinct feature of these findings is the discovery that the closer to zero energy a molecular negative ion state is, the larger the ability of molecules to remove electrons from the medium by attachment. Science revealed that *the ability of molecules to capture slow electrons increases as the energy where these transient anions are formed is lowered; slow electrons are captured by molecules much more efficiently than fast electrons are.* In fact, Figure A. 3 indicates that as the position of the negative ion states of molecules approaches thermal energy, the

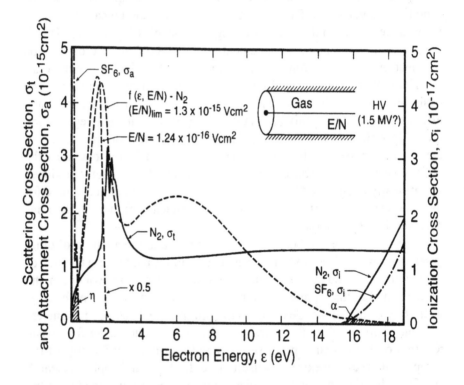

Fig. A. 2. Total ionization cross section $\sigma_i\,(\varepsilon)$ for N_2 (__) and SF_6 (-.-) close to the ionization threshold. Total electron scattering cross section $\sigma_t(\varepsilon)$ as a function of electron energy, ε, for N_2 (__), and total electron attachment cross section $\sigma_a\,(\varepsilon)$ for SF_6 (-.-). Electron energy distribution functions f (ε, *E/N)* in pure N_2 for two values of *E/N* : at 1.24 x 10^{-16} V cm^2 -- this value of *E/N* is about ten times lower than the *E/N* value at which breakdown occurs under a uniform electric field -- and at the limiting value of *E/N* (1.3 x 10^{-15} V cm^2) at which breakdown occurs under a uniform electric field. The shaded areas designated by η and α are, respectively, a measure of the electron attachment and electron-impact ionization coefficients for SF_6 (from Christophorou[8]).

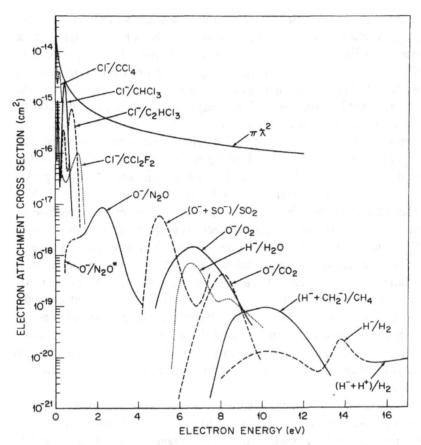

Fig. A. 3. Dissociative electron attachment cross sections as a function of the electron energy for a number of molecules. Parent negative ions are also formed at the extreme low-energy range [e.g., SF_6^-; the attachment cross section for SF_6 is shown in Fig. A. 2 and is mostly due to the formation of SF_6^-]. The curve designated by $\pi(\lambda/2\pi)^2$ represents an upper limit to the electron-capture cross section.[9]

probability of electron capture by certain molecules becomes almost equal to one: every electron that encounters such molecules is captured by them. Many molecules rupture easily in such encounters and decompose revealing their extreme fragility in their encounters with slow electrons.

Knowledge on Reactions (3) along with knowledge on Reactions (1) and (2) has helped the development of gaseous insulation for the transmission, distribution, and conservation of electric power. Until recently the most common insulation medium for high-voltage transmission of electricity was air. Because air inhibits breakdown only if the high-voltage conductor is a long distance from other objects, high-voltage transmission lines are strung between towers and tall poles to separate them sufficiently from the ground. Insulating power lines this way requires considerable land, is not aesthetically pleasing, allows emission to the environment of low-frequency electromagnetic radiation, generates corona noise, and since there is a limit as to how tall the poles can be made, there is a limit on the size of the transmission voltage that a transmission line can be operated at for a given level of electric power. In the United States, energy losses caused by electrical resistance in such electric power transmission lines amount to 5 % to 10 % of the electrical energy transmitted this way. Many of these disadvantages can be overcome by using alternative (to air) gaseous dielectrics such as sulphur hexafluoride (SF_6) to insulate electric power transmission lines and other equipment such as substations, circuit breakers, and transformers. Because these insulating gaseous materials have superior high-voltage properties than air, they allow utilities to employ more compact equipment and enclosed underground cables that are aesthetically more acceptable and can operate at higher voltages. Less land is needed for these cables, the cables are hidden, electromagnetic radiation can be eliminated, higher transmission voltages can save energy, and substations can be installed in cities close to the loads.

Compressed-gas cables have been in service since 1968. Gas-insulated circuit breakers are in common use and gas-insulated transformers is an emerging technology with distinct advantages (e.g., with regard to flammability) over oil-insulated transformers. Gas-insulated equipment is now a major component of the power transmission and distribution systems all over the world and it employs almost exclusively the man-made gaseous dielectric SF_6.

Let us then see how basic scientific knowledge has helped developed such gaseous insulators for the electric power industry. In the inset of Figure A.2 is schematically depicted an enclosed gas-insulated transmission line. In practice, the conductor is held in place in the center of the pipe by solid insulators and the pipe is at ground potential. The question to be answered is this: What gas should be used to insulate the high-voltage transmission line (or lines) and at what pressure so that for reasonable size pipe the power can be transmitted at voltages, say, in excess of 500,000 Volts? Rephrasing the question: What are the fundamental processes that make a good high voltage gaseous insulator? To answer this question we may first ask. Why does the gas fail to insulate the high voltage transmission line when the voltage is beyond a given level? How can one identify a gaseous medium consisting of one or more components which is a relatively poor conductor or a nonconductor of electricity to very high applied voltages?

The answer to these questions can be formulated by referring to Figure A.2 which links the basic physical processes that underline the development of high-voltage gaseous insulation to be used for the technology schematically represented by the inset in the figure. When there is no voltage on the conductor the gas is naturally a perfect insulator. When the gas is electrically stressed, the free electrons present[10] in the gas gain energy from the applied electric field. At a certain level of the applied voltage an appreciable number of these free electrons has kinetic energies high enough to ionize the gas molecules (Reaction 2) and multiply rapidly in an avalanche-like fashion so that the gas breaks down (i.e., the gas makes the transition from an insulator to conductor). The minimum critical voltage under which the electrical conductivity of the gas increases by many orders of magnitude is known as the breakdown voltage, and the phenomenon as electrical breakdown, or electrical discharge, or spark.

While a multiplicity of physical processes involving electrons, photons, and ions play a role in determining the breakdown strength and other dielectric properties of a gas, basic research has shown that the slow electron is the key particle and its interactions with the gas molecules are the critical processes, especially those processes which produce (Reactions 2), deplete (Reactions 3b and 3c), or slow down (Reactions 1 and 3a) free electrons.[8] Basic knowledge on these processes and measurement of their corresponding cross sections allowed a prediction of the dielectric properties of gases and the choice of appropriate gaseous media for specific uses by the electric power industry. To see how this has been accomplished let us return to Figure A. 2. In a gas -- at a number density N -- under an applied electric field E, the free electrons present in it attain an equilibrium energy distribution $f(\varepsilon, E/N)$ which is a function of the gas and the ratio E/N. When the value of E/N is low, the electron energy distribution function $f(\varepsilon, E/N)$ lies at low energies and the number of electrons capable of ionizing the gas is negligible (i.e., the gas is an insulator). As the voltage is increased, $f(\varepsilon, E/N)$ shifts to higher energies. For sufficiently high E/N, the number of electrons with energies high enough to ionize the gas molecules is such that the gas makes the transition from an insulator to a conductor (i.e., it breaks down). Even at the E/N value at which breakdown occurs, only a small fraction of the electrons present in the gas have sufficient energy to induce ionization. This is schematically shown in Figure A. 2 by the shaded area designated by the Greek letter α (a measure of the so-called ionization coefficient of the gas). For a non-electron attaching gas such as N_2, knowledge of α/N as a function of E/N provides a measure of the breakdown field. For an electronegative gas such as SF_6 which attaches free electrons the situation is different. In this case the free electrons are captured by the gas molecules forming stable negative ions and are thus prevented from initiating breakdown. As can be seen from Figure A. 2 the cross section for electron attachment to SF_6 is large at very low energies and thus only electrons with energies at the extreme low-energy range can be removed efficiently from the dielectric by electron attachment. The shaded area in Fig. A. 2 designated by the Greek letter η schematically represents a measure of the electron attachment coefficient η/N of the gas. Knowledge of α/N and η/N as a function of E/N allows prediction of the value of E/N -- that is, for a given pressure and system dimensions, the value of the applied voltage -- at which the gas insulation breaks down. For uniform electric fields this is the value of E/N at which $\alpha/N = \eta/N$.

The dielectric strength of a gaseous insulator can be optimized by using basic knowledge on the ionization coefficient α/N (E/N) (which must be small), the electron attachment coefficient η/N (E/N) (which must be large), and the electron energy distribution $f(\varepsilon, E/N)$ (which must be shifted to low energies to minimize electron production and to maximize electron removal by attachment). Many basic studies have shown that high breakdown strengths require gaseous media with large electron attachment cross sections. It is thus apparent that the insulating properties of gases can be optimized by a combination of two or more gas components designed, for example, to provide the best effective combination of electron attaching and electron slowing-down gases. Knowledge of the electron attachment cross sections of many gases not only led to the discovery of many excellent insulating gases (such as SF_6), but also to the appropriate choice of electronegative components for dielectric gas mixtures. Similar knowledge on electron scattering at low energies guided the choice of buffer gases for mixtures containing electronegative gas additives. Of unique practical significance are mixtures of the strongly electron attaching gas SF_6 with the inert, inexpensive, and environmentally acceptable gas N_2 with which it acts synergistically: the nonelectronegative gas N_2 scatters electrons (principally through the indirect negative ion state at 2.3 eV) into the low-energy region where the electronegative gas SF_6 removes them efficiently by attachment. At the present time such mixtures are in limited use, but they are anticipated to become more attractive in the future.

Recently, this research area has assumed a particularly significant importance because the widely used industrial insulating gas SF_6 has been found to be a potent greenhouse gas and thus the need has arisen for gaseous insulation substitutes that are environmentally acceptable. The N_2/SF_6 gas mixtures designed on the basis of fundamental collision data offer a viable alternative.[11]

References and Notes for Appendix C

1. An electron with 100 eV energy has a speed of about 6 x 10^8 cm/s . An electron in thermal equilibrium with its surroundings at room temperature (about 300 K) has an average kinetic energy of about 0.04 eV and is 50 times slower (its average speed is about 1.2 x 10^7 cm/s). An 100 eV electron has a de Broglie wavelength of about 1.2 Å while an 0.04 eV electron has a de Broglie wavelength of about 60 Å. The slower the electron the longer its de Broglie wavelength and the larger the probability (cross section) of interaction with the atoms or molecules it encounters.
2. L. G. Christophorou, *Atomic and Molecular Radiation Physics*, Wiley-Interscience, New York, 1971.
3. L. G. Christophorou (Ed.), *Electron-Molecule Interactions and Their Applications*, Academic Press, New York, Vols. 1 and 2, 1984.
4. Process (ii) is especially significant when laser light is used to increase (by photodetachment of negative ions), or to deplete (by attachment to laser-excited molecules) the number of electrons present in the system.
5. C. K. N. Patel, Phys. Rev. A **136**, 1187 (1964).
6. The CO_2 molecule has three fundamental vibrational modes: the symmetric stretch (v_1, 0.17 eV spacing), bending (v_2, 0.08 eV spacing), and asymmetric stretch (v_3, 0.29 eV spacing). The number of quanta in each mode are given by the numbers $v_1 v_2 v_3$. The notation for a particular vibrational level is $v_1 v_2^\wedge v_3$, where $\wedge$ indicates the electronic orbital angular momentum of the molecular state.
7. A. J. DeMaria, Proc. IEEE **61**, 731 (1973); Ref. 3, Vol. 2, p. 325.
8. L. G. Christophorou, *Insulating Gases*, Nuclear Instruments and Methods in Physics Research A **268**, 424 (1988).
9. Ref. 3, Vol. 1, p. 559 .
10. These electrons come from various terrestrial and extraterrestrial sources, from thermionic emission, and from electron detachment from negative ions.
11. L. G. Christophorou, J. K. Olthoff, and D. S. Green, *Gases for Electrical and Arc Interruption: Possible Present and Future Alternatives to Pure SF_6*. National Institute of Standards and Technology Technical Note 1425, 1997.

Index